£3-75

THE MATTER OF LIFE

The Matter of Life PHILOSOPHICAL

PROBLEMS OF BIOLOGY   BY MICHAEL A. SIMON

New Haven and London
Yale University Press
1971

Published with assistance from the foundation
established in memory of Henry Weldon Barnes
of the Class of 1882, Yale College.

Library of Congress catalog card number: 74-158142
International standard book number: 0-300-01500-3

Designed by Sally Sullivan
and set in IBM Selectric Baskerville type.
Printed in the United States of America by
The Carl Purington Rollins Printing-Office
of the Yale University Press.

Distributed in Great Britain, Europe, and Africa by
Yale University Press, Ltd., London; in Canada by
McGill-Queen's University Press, Montreal; in Mexico
by Centro Interamericano de Libros Académicos,
Mexico City; in Central and South America by Kaiman
& Polon, Inc., New York City; in Australasia by
Australia and New Zealand Book Co., Pty., Ltd.,
Artarmon, New South Wales; in India by UBS Publishers'
Distributors Pvt., Ltd., Delhi; in Japan by John
Weatherhill, Inc., Tokyo.

315664

For Shannon

Contents

Preface

If scientific knowledge has a prototype, it is the sort of knowledge afforded by physical science with respect to inanimate nature. It is this domain that yields most readily to scientific understanding, in contrast to the realm of man and his social groupings, which seems to be much less tractable. Living systems, the subject matter of biology, appear to occupy a middle ground: they represent aggregates of matter for which an elaborate science has been contrived, albeit one that lacks both the scope and formal impressiveness of the physical sciences. Someone concerned with the nature of biological science, accordingly, may wish to ask to what extent a biological science based on the physical paradigm is possible, and how the form of knowledge that comprises that science is determined by the peculiar character of its subject matter. He may also inquire as to what biological science presupposes, what it can tell us that physical science cannot, and whether there is anything about living things that neither physical science nor biological science can deliver. These are the main questions for which the study that follows attempts to provide some answers.

At the time I began this work, my interest was in trying to discover what, if anything, is distinctive about biological science, its concepts, and its mode of explaining. I wanted to know what it is for something to be an organism, to be alive, and what sort of science could provide an intellectually satisfying answer to that question. The path of exploration has taken me to consider the broader questions of the relations among the sciences in general, and of the relation between a science and that which it purports to elucidate. What has emerged, I believe, has bearing not only on the question of the nature of science, but also on the question of the relation of science to man. Biology differs from physical science because its objects are different. A man is different from a stone, and the fact that biology can deal with the former but not the latter tells us something both about man and about biology. The science of life is not the same thing as a science of man, nor can it be expected to provide all that a proper understanding of man may demand. To understand biological science, and what it can and cannot do, however, may represent a significant step toward understanding not only one of man's intellectual products, but perhaps his nature as well.

I have benefited at all stages of this work from help and support from friends, colleagues, and associates. To my colleagues at the University of Connecticut, especially, but by no means exclusively, those in the philosophy department, I am grateful for ideas and suggestions, as well as for much spirited and trenchant criticism. The contributions of those philosophers, biologists, and social scientists, though too numerous to list, will not go unremembered. In addition, I wish to thank the University of Connecticut for its material support, both through its Research Foundation and its Summer Faculty Fellowship program; Helen Rollins, for her exemplary proofreading; and Jane Isay of Yale University Press, for the conscientiousness and enthusiasm with which she has mediated the transformation of manuscript to book. But most of all, I am indebted to my wife, Shannon, who, more than anyone else, has helped me to appreciate what else besides a biological organism a human being is.

M. A. S.

Storrs, Connecticut
March 1971

Introduction

The world contains, among other things, material bodies whose structure and behavior have earned them the designation of living. For Aristotle the difference between animate and inanimate objects consisted in the possession of *psyche,* the principle of life. Biology is the science which attempts to define and elucidate that principle and to render intelligible the things that have it. Biological inquiry is concerned with how they are identified, how they reveal their distinctiveness, how they do what they do, and how they have come to be the way they are.

The diverse strands of the fabric of modern biology may be seen to represent the course of man's efforts to provide answers to these various questions, strands which have come to realize a greater integration as the science has gained in maturity. Taxonomy provides a systematic basis for recognizing and classifying biological entities according to data provided by anatomy, physiology, and biochemistry. Understanding the means by which living things perform as they do is served not only by anatomy (including microanatomy), physiology, biochemistry, and biophysics, but also by systems theory and the information sciences. Ethology and ecology extend further the range of biological knowledge by taking in molar behavior and relations to features of the world which fall outside the boundary of the individual organism. Genetics and embryology, finally, offer accounts of the various features of biological individuals as products of historical processes, both with respect to inheritance and with respect to the development of the individual—processes whose mechanisms have become objects of study for the molecular sciences.

Since Darwin, biology as a historical subject has displayed an additional explanatory aspect. In order to explain the individual's coming to possess a particular inherited characteristic, the biologist is concerned to show not only how this feature was passed on from the creature's biological forebears and how it developed from embryonic to mature form within the single organism, but also why the individual and others who possess it have this particular characteristic rather than some other. In other words, the question becomes one of how the *earth* has come to contain creatures with character X rather than character Y; that is the question for which the theory of evolution by natural selection provides an answer. Thus extended, biological science can be construed as concerned not merely with organisms but with certain features of the earth or even of planetary objects in general.

There are a number of ways of dividing up the class of activities that fall under the heading of biological research. One is according to the types of organisms studied; thus there have been microbiologists, vertebrate zoologists, invertebrate zoologists, entomologists, botanists, mycologists, and so on. Another mode of division is according to the biological processes investigated, regardless of the organisms in which they occur; classical genetics and embryology, as well as the various subdivisions of physiology, such as endocrinology and neurophysiology, are examples of this type of division. Most biologists today, however, as P. B. Medawar has pointed out (*88*, p. 101), tend to classify themselves by the analytical levels at which they work, rather than by the subjects they are investigating. Medawar distinguishes four levels or strata, occupied by the molecular biologists, the cellular biologists, biologists who work at the level of whole organisms, and biologists who study populations of organisms. Since it is a characteristic of workers at any scholarly or creative enterprise to regard their own work as the most important, it is not surprising that it is over the relations between these levels that many of the issues concerning what biology really is (or should be) arise. In particular, biologists have disputed over whether the phenomena of life can be accounted for in terms of that which is nonliving. Can a biological system best be understood as an evolving, self-replicating aggregate of molecules, or must we say that the matter of life has properties beyond those possessed by matter in general? Can biology explain what it is for an organism to be alive, while abstracting from the fact that it is alive?

The dispute is as old as Greek philosophy. Prior to the present century, it was simply the controversy between mechanism and vitalism, whether life processes are to be held ultimately explicable in the same terms as are adequate for inanimate matter or whether it is necessary to assume the operation of nonphysical factors or vital principles in order to explain the phenomena of living systems. By the 1920s mechanism versus vitalism had essentially become a dead issue, but it meanwhile had given rise to a new controversy: the dispute between mechanistic and "organismic" biology. According to the organismic biologists, biology can be understood only in terms of holistic notions and must employ concepts and assert relations that are neither defined in nor derived from the physical sciences—claims which were typically denied by mechanists. The contemporary version of these disputes is concerned with whether biological organisms are explicable in terms of chemistry, and whether biology is an autonomous discipline or merely an application of chemistry and physics. The issue that persists is over whether biological phenomena have to be reducible in order to be intelligible.

The central philosophical problem concerning the nature of biological science is that of understanding the relation between biology, viewed both as an activity and as a set of propositions, and the world of nature from which that science is drawn. It is the problem of determining and assessing the ways in which the special demands posed by biology's peculiar subject matter are revealed in the methodology and the conceptual apparatus of that discipline. It is the problem of ascertaining how and to what extent the characteristics of living organisms determine a science that is significantly different from the other natural sciences.

In the study which follows I shall attempt to approach this problem from a number of perspectives. One of these will focus on the question of biology's relation to the physical sciences and its status as a science in its own right. Another will be concerned with the types of explanations used in biology and their logic. Yet another will involve consideration of the nature and the role of theoretical models in biology, with particular reference to the gene, and will include a detailed case history of that construct. Assessing the presuppositions of biological science will constitute still another means of approaching biology's distinctiveness. Examination of the limits and implications of extending biological concepts to other areas of thought and action, especially with regard to human concerns, will represent a final attempt to grasp biology's relation to the world. Philosophical implications of biological science are to be found on a variety of levels of scrutiny.

Biology owes its special interest both to the fact that it deals with what traditionally has been the more enigmatic side of the animate-inanimate distinction, and to the fact that it is about, among other things, ourselves. The unit of personal identity is also the basic unit of biological organization, the individual organism. Responses to the results of scientific inquiry typically exhibit two opposing attitudes: wondrous awe at finding something apparently inexplicable in terms of anything else, and aesthetic pleasure at having something neatly explained in terms of recognized general principles or fundamental intelligible types. The conflict between these attitudes with respect to biology has been especially acute, perhaps because biologists are persons and persons are organisms. In any event, the elucidation of biological systems—or rather the very possibility that there be such elucidation——provides a subject whose philosophical dimensions range well beyond those of science in general.

1. Biology as Science

If the existence of a science of biology is the result of man's efforts to understand living organisms, then the existence of a science of physics is the result of his efforts to understand certain phenomena of inanimate nature. Seen this way, the biological and the physical sciences bear similar relations to their respective domains. On the other hand, there are also a number of striking dissimilarities between biology and physics, among them biology's large descriptive component, its apparent failure to exhibit universal laws and theoretical principles, and the asymmetry of its relation to physics and chemistry: the fact that the applicability of physicochemical principles to biological subject matter is nonreciprocal. Physics is clearly more basic than biology; is it thereby more scientific?

There are a number of different approaches to the problem of what kind of science a science of life actually is. One of these involves the comparison of its subject matter with that of the other sciences. Another concerns the nature of its formal structure and its descriptive and explanatory apparatus. Still another requires an examination of the conditions and the possibilities for translating its statements into those of another science. Understanding the nature of biological science both contributes to and is augmented by the study of the relations between the sciences. Both of these interests may be served by a consideration of the relations of that science's propositions to its subject matter, to each other, and to the propositions of the other sciences.

Biology, Physical Science, and Natural History

According to a familiar conception of the nature of the sciences, there exists a fundamental division between the descriptive and the explanatory sciences. The descriptive sciences, it is sometimes maintained, simply take the phenomena as they find them and arrange them into classes according to their observable characteristics. Explanatory science, on the other hand, is supposed to attempt to find the *basis* for whatever regularities are discovered. The distinction, thus understood, is between representations of what there is and representations of the ways these things hang together.

Defenders of the distinction are usually willing to concede, however, that every science has both descriptive and explanatory aspects. It will thus be pointed out that physics offers, among other things, a catalog of the fundamental particles of matter, and that chemistry delivers descriptions of the properties of various chemical substances, as well as attempting to account for their relationships in terms of underlying structure. A descriptive science like geology, on the other hand, may provide explanations of the origins of lakes and mountain ranges, and physiology may explain the working of hearts and kidneys. Nevertheless, it may still be maintained that a sharp division exists between physical science and natural history, with the latter including the descriptive sciences. Whereas natural history is supposed to be concerned only with particular facts about the physical universe in which we happen to live, physical science is understood to deal with the formal principles in accordance with which our galaxy, and every other galaxy, is fundamentally structured. Natural history offers an index of the regularities to be found in the world, plus an account of the sequence of states and events that have preceded the present state of things. Geology, astronomy, and biology are, accordingly, branches of natural history because of their concern with describing sets of naturally occurring objects and relating their histories. They are considered branches of physical science only to the extent that they offer explanations in terms of universal laws of nature.

Applied to biology, this view amounts to the assertion that biology, so far as it is a science at all, is ultimately biochemistry and biophysics. Anatomy, taxonomy, embryology, physiology, and genetics all are branches of natural history except as they deliver elucidations of the processes of biology on the level of chemistry and physics. Organs, cells, and intracellular parts, such as chromosomes, mitochondria, and even DNA molecules, are logically on a par with ferns and ferrets: terrestrial objects to be described and made subject to empirical generalization. According to one clear statement of this position, that of J. J. C. Smart (*134*, pp. 92-96), biology, because it is concerned with the properties of merely certain classes of highly complex objects which happen to be found on the surface of one particular planet, cannot have laws, in any strict sense, but only generalizations which, as part of the natural history of the earth, may quite possibly be falsified either by subsequent discoveries on our own planet or by as yet unknown phenomena of extraterrestrial natural history. Only physics and chemistry are presumed to provide laws that apply everywhere in space and time. Biology can be other than natural history only to the extent that its explanations are ultimately biochemical and biophysical. And there will be no lawlike principles of biochemistry and biophysics that are not themselves principles of chemistry and physics proper.

The effect of thus regarding the biological sciences as essentially physics and chemistry plus natural history is to drive an acute methodological wedge between physics and biology. Biology is not presumed to be a subject of the same logical sort as physics; rather, it is conceived of as the study of a certain class of naturally occurring objects, and of the application of chemistry and physics to them. Thus Smart has suggested (*135*, p.57) that biology is related to physics and chemistry in the way in which radio engineering is related to the theory of electromagnetism: physics and chemistry are used to explain how organisms carry on their natural functions, in the way that physics is used to explain how an electronic instrument such as a television set works. Of course, the biologist, unlike the electronics engineer, is not presumed to regard his subject matter as a portion of applied science. His interest is rather in the natural history of structure and in why things with this structure behave as they do.

A scientist who supports this interpretation of biology

is committed to look upon organisms as essentially highly complicated machines, as physical mechanisms whose structures and processes are ultimately explicable in terms of physics and chemistry. Thus conceived, organisms would no more be expected to be the subjects of laws and theories than would automobiles or typewriters. Empirical generalizations may be possible, but these would not have the status of laws of nature, any more than would generalizations about the ways existing types of automobiles or typewriters work. If biology is construed as basically the study of the ways that certain machines which happen to occur in nature are constructed, then it can hardly be said to exist as an autonomous science, except as a species of natural history.

The justification and other implications of regarding organisms as physical mechanisms will be considered in a later chapter. An important question that remains, however, is whether viewing biology as essentially natural history of structure really differentiates it from chemistry and physics. Indeed, there may be a sense in which virtually all science is natural history, and the differences between the sciences may be considered to consist merely in the fact that they deal with hunks of stuff of different sizes and different natures. As astronomy studies stars, geology studies the inorganic earth, and biology studies the organic earth, so chemistry studies molecules, and physics studies atoms, particles, and forms of energy. Is any of these sciences truly ultimate or fundamental in an absolute sense? Perhaps not. Perhaps all of them should be construed as basically the "natural history" of the things and combinations of things that happen to be found in the portion of the universe that (terrestrial) man has investigated.

One might argue, for example, that the science of

chemistry, in its role of characterizing the various chemical compounds that are found on or near our planet, is really the natural history of certain aggregates of matter that we have happened upon. Even the elements are aggregates, being composed of individual atoms, which are themselves configurations of subatomic particles. Certainly a great deal of chemical research is concerned with investigating the structure and behavior of natural products, whether they be proteins, boron hydrides, or alkaline earth elements. It may be somewhat less than convincing to maintain that the study of seaweed, cells, and chromosomes is natural history, while insisting that the study of carbohydrates, hydrocarbons, and oxides of chromium is not.

In defense of regarding the laws of chemistry as distinct from propositions of natural history, it might be replied that chemistry, in principle, treats the behavior of *all* aggregates of the fundamental physical particles that are possible according to the laws of physics, whereas biology is concerned essentially with only that minute fraction of possible "biological entities" that happen to inhabit the earth. Biology has nothing that corresponds to the Periodic Table, every one of whose places is occupied by a known substance. Furthermore, chemistry is concerned not only with actual elements and compounds, but also with possible (and impossible) compounds, compounds whose existence (or nonexistence) and properties can be predicted on the basis of the physical principles that determine what is possible in the universe at all. Chemistry thus seems to possess a kind of universality that biology lacks.

What this argument shows, however, is not that chemistry is not basically concerned with the natural history of material aggregates, but only that it has achieved considerable success in producing as a result of its investigations a number of theoretical principles that have allowed chemists to proceed *as if* chemistry is truly universal, something biology has not done. Chemical taxonomy has "slots" because chemistry has theory. But the existence of a Periodic Table was obviously not necessary for the chemical elements and compounds to have been characterized, that is, for "chemical natural history" to have been written. Chemistry did not suddenly pass from the status of natural history to that of physical science as a result of Mendeléef's work.

The fact that chemistry provides us with a theoretical basis for predicting what sorts of things can and cannot be expected to exist does not adequately distinguish it from biology. In the first place, we have no way of excluding the possibility that there may be some limits to biological possibility that we do not know about. Secondly, even now biology can specify important boundary conditions beyond which particular processes are known either not to occur at all or to be altered in predictable ways. Investigations of the effects on mammalian physiology of sustained periods of weightlessness, for example, represent attempts to determine such limiting conditions. After all, the constraints on chemical reality that we speak of have only been established as part of physicochemical theory, which is based on what we find and what we do not find. The difference between chemistry and biology, therefore, would appear to be the difference not between physical theory and natural history but rather between natural history with theory and natural history without theory.

According to the view that draws a sharp distinction between physical science and natural history, it is the supposed universality of the laws of chemistry and phys-

ics that sets them apart from the principles of biology
and those of the other terrestrial sciences. The laws of
physics, such as those of classical mechanics and electro-
dynamics and the equations of quantum mechanics, and
the laws of chemistry, expressed by chemical equations,
are assumed to be universal in that they are supposed to
"apply everywhere in space and time, and . . . be expressed in perfectly general terms without making use of
proper names or of tacit reference to proper names"
(*135,* p. 53). These are contrasted with biological propositions, all of which are considered to carry implicit reference to our particular planet, since all biological concepts must ultimately be defined in terms that make
implicit reference to the particular entity *earth.* If we
try to define biological terms without reference to the
earth, so it is argued, then there is no reason to suppose
our biological propositions to be universally true. Somewhere in the universe, perhaps in some other galaxy, a
counterinstance is likely to be found. Only the laws of
chemistry and physics can be presumed to admit of no
exceptions.

In support of this account of the relative status of
biological and physicochemical principles, it must be
acknowledged that the former do indeed seem to be
earthbound in a way that the latter do not. From a
strictly logical point of view, however, the distinction
may be challenged. Although it is true that a proposition such as "All mules are sterile" or "Copper deficiency produces anemia" is limited in its applicability, its
failure to preclude exceptions when expressed in strictly
nonterrestrial terminology may be considered simply a
consequence of the fact that biological principles are
not ordinarily spelled out in such a way as to indicate
their scope and limits. It is generally assumed by biolo-

gists that if all of the relevant features of both the organism and environment were specified, the propositions of biology would be as free from exception as those of chemistry and physics. Biological concepts are "open-textured": they are not strictly and antecedently defined in such a way as to leave no doubt as to their applicability or nonapplicability to as yet undiscovered instances. To construe an apparent counterinstance as falsifying a biological principle is to fail to recognize this open-textured quality. An example of such an error would be to regard marsupial mammals as falsifying certain propositions made concerning mammals in general. Only if it is impossible to attribute the differences in the behavior of the alleged counterinstances to some additional feature with respect to which they differ from the "standard" cases will it ordinarily be proper to say that the original generalization has been refuted.

What is crucial here is that all of these considerations apply to chemistry and physics as well, albeit to a less conspicuous degree. Thus relativistic phenomena are usually regarded not as falsifying classical mechanics, but only as showing that Newton's laws of motion can be expected to hold only under certain limiting conditions, conditions that were not mentioned at all in the traditional formulations of the laws. Similarly, the principle of the immutability of the chemical elements may be seen as yielding not to refutation but merely to qualification as a result of the development of atomic theory. Another illustration is provided by the discovery of stereoisomerism in chemistry, a discovery that demanded recognition of a difference between alternative forms of chemical substances possessing the same molecular and structural formula. To declare any science universal is only to presume that there do not exist unknown phenomena the discovery of which would require one to limit further the scope of existing theories. It may be intuitively reasonable to suppose that chemistry and physics have an immunity to subsequent revision which far outstrips that of biological principles. But simply as a point of logic, there can be no more basis for calling chemistry and physics truly universal than there can for biology.

Part of the basis upon which physics (and to a lesser extent chemistry) has been regarded as possessing a universality not shared by the other sciences is the allegation that it deals with "simples," or else with things that are homogeneous. Thus Smart (*135*, p. 55) points to the fact that classical particle mechanics, which forms the basis for rigid mechanics, deals with point masses, and that the physical properties of the atom can be explained because the theory of the atom is reducible to that of simpler particles which are taken to be ubiquitous in the universe. Macroscopic laws arise on account of the statistical behavior of submicroscopic particles, so that a gas can be treated as homogeneous in a way that organisms, being "vastly complicated and idiosyncratic structures," cannot. Biology is thus inevitably to be afflicted with exceptions owing to the complexity of the structures with which it deals.

As Smart recognizes, however, what we have here is only *relative* simplicity and homogeneity. These features cannot therefore be expected to provide the basis of a radical distinction between the biological and the physical sciences. What we are faced with is a continuum of cases: frogs may be more complicated than bacteria, blood serum less homogeneous than water, and protein molecules more complicated than carbon atoms, which are less simple than electrons. Each science is concerned with a variety of things some of which are simpler and more homogeneous than others, but even physics deals with properties of things that are not necessarily assumed to be absolutely simple or homogeneous.

It also deserves to be pointed out that much of physics is not concerned with structure anyway, so the role of simples and the homogeneous may be quite different from what is sometimes supposed. Classical mechanics and electromagnetic theory, for example, can be construed as studying the relations between certain properties that happen to attach to physical systems. The "objects" whose behavior is described in these physical laws are merely theoretical fictions that are neither observed nor inferred. There are no such things as point-masses and perfect conductors and ideal gases. The theory of the behavior of these nonentities is not physics, any more than the theory of the behavior of a utopian society is sociology. Physics, like every other science, is about real objects, and real objects are heterogeneous. It is not a branch of mathematics, like geometry. If the propositions of physics are true, it is because the heterogeneous things that exist in the world behave in ways that justify the application of the concepts of uninterpreted physical theory to them. By the same token, biology is about real objects—namely, those that exhibit the properties that qualify them to be classed as organisms—and contains true propositions only to the extent that these objects display the characteristics that make these propositions applicable to them. There is no science for which there cannot be conceived a distinction between uninterpreted and interpreted, or between pure and empirical.

The discussion thus far has pointed to the conclusion that the sciences all bear essentially similar relations to

their subject matter. Every science is properly understood as being *about* something, even if it is only because we tend to hypostatize whatever it is we happen
to be studying, whether it be atomic nuclei, heat trans-
fer, or electromagnetic radiation. Some sciences may be
considered more basic or fundamental than others, in
the sense that they deal with stuff some of which forms
the components of other kinds of stuff which constitutes the domain of other sciences, but it does not follow that any one is absolutely fundamental, or that the
less basic ones are any less "scientific" than the more
basic ones.

Given the point that all science is ultimately natural
history and that biology is a bona fide science on all
fours with the others, some biologists, wishing to stress
the breadth and variety of their subject, have appeared
to suggest that the biological sciences are, if not more
basic or fundamental, more general than the physical
sciences. Thus C. F. A. Pantin has argued that the fundamental contrast between the biological and the physical
sciences lies in the fact that the former are *unrestricted*
sciences, whereas the latter are *restricted* (*103,* chap. 1).
In physics and chemistry, investigators restrict their attention to lower levels of organization, thereby excluding (it is argued) much of the wealth of natural phenomena. The investigator in biology or geology, on the other
hand, is prepared to follow the analysis of his problems
into any other science whatever.

A somewhat similar position has been put forward by
G. G. Simpson, who has attempted to show that biology
"is the science that stands at the center of all science"
(*131,* pp. 106-07). The principle that allows Simpson to
reach this conclusion is to be found in his suggestion
that the characterization of science as a whole should be

sought not by looking for principles that apply to all phenomena, but by looking for phenomena to which all principles apply. Since all known material processes and explanatory principles allegedly apply to organisms, while only a limited number of them apply to nonliving systems, biology's central place among the sciences is assured.

If we take Simpson's position to its logical extreme, however, we may be forced to give the central slot not to biology but to social science, since there seem to be principles which apply to social groups but not to individual organisms. Simpson's concept of the centrality of a science thus appears distinctly Pickwickian, if not blatantly anthropocentric. Presumably he would want to say that it is only by virtue of the fact that the central science includes the principles that concern lower levels of organization that it enjoys its privileged status. If this means that biology is taken to *include* chemistry and physics (whereas the social sciences obviously do not), then perhaps the most Simpson can be accused of is having delusions of grandeur.

Pantin's position, though more moderate, also suffers a certain awkwardness resulting from appearing to have generality attach to phenomena rather than to propositions. The point is not to include chemistry and physics within biology, but rather to note that biology includes in its subject matter phenomena which are also the subject matter of the physical sciences. It is not the principles of these sciences that are restricted, but their domains, the levels of organization to which they apply. What might make Pantin's view seem paradoxical is that it requires that the restricted sciences be those that involve levels of organization that are to be found in *all* things. But there is no way of evading the fact that the

unrestricted sciences, whether biology, geology, meteo-
rology, or nuclear physics, possess that feature only because they are specialized with respect to their basic
subject matter.

Every science is specifically concerned either with certain classes of material things or with certain classes of
properties. In many cases we are more directly interested in the properties than we are in the sorts of things we
actually study when we carry on research. Some sciences, notably chemistry and physics, have sets of properties which they take as their own; thus we commonly
distinguish between the chemical and the physical properties of a substance. But do we ever speak in this way
of biological properties or of geological or astronomical
properties? Apparently not. The reason seems to be that
whereas physical and chemical properties are present in
anything whatever, astronomical and biological properties would properly be said to be present only in systems specifically and antecedently recognized as astronomical or biological objects. A biological property can
be defined only as a property peculiar to biological systems; applied to anything else, it would be taken as at
best metaphorical, at worst totally inappropriate.

It must be recognized, of course, that what constitutes
a thing as a "biological object" is the possession of biological properties, just as it is the possession of chemical
or physical properties that makes something a "chemical
object" or a "physical object." However—and here is the
point—anything at all will be said to have chemical and
physical properties, including those things that are attributed astronomical or biological properties. In other
words, chemistry and physics concern the most general
properties, and the other sciences concern less general
ones. What we have is a hierarchy of sciences, based on

degree of generality. Thus chemistry, which deals with all matter, is less general than physics, which is not restricted to studying matter, and more general than biology and astronomy, each of which deals only with certain complexes of matter. (When it comes to the social sciences the situation is more complicated, although it can still be maintained, for example, that psychology, since it concerns a subclass of organisms, is less general than biology; sociology may then be construed as still less general, since not all animals that are capable of being subject matter for psychology belong to social groups.) Biology thus conceived would then be not an application of chemistry and physics but rather a complementary set of propositions for describing the natural world. It is a ramification not of physics but of science in general.

Biology's Autonomy and the Problem of Reduction

Biology is a specialized subject in the sense that its predicates are applicable only to organized wholes and that these wholes comprise a relatively small subset of all of the wholes in the universe. Does it stand alone, with its own set of propositions, or is it in some sense reducible to the more general sciences, chemistry and physics? Biology is the science of biological objects; is such a science an autonomous discipline?

Whether or not biology is to be considered autonomous will, of course, depend on what we are to mean by "autonomous." If this term is taken to indicate, for example, that living systems are composed, at least in part, of irreducibly biological components, such as primordial germ-particles or nonmaterial elements of a spiritual nature, then the verdict of virtually all modern biologists (at least qua biologists) will be that biology is definitely not autonomous. Vitalism of a substantive type is simply not a position that one will find defended by contemporary biologists. Biology is not to be considered autonomous in the sense of possessing a subject matter that is materially discontinuous with that of other natural sciences.

On the other hand, if biological autonomy means merely that the complex components of biological generalizations have not yet been translated into physicochemical terms, then biology surely is an autonomous science. Though it may be commonly believed that all concrete biological entities are ultimately composed of the same basic ingredients as the nonorganic parts of the world, this belief by no means contravenes the autonomy of biology, assuming this is taken to mean only that biology employs concepts and asserts relations that are neither defined in nor derived from the physical sciences. The possibility of a science of classical genetics or of thermodynamics in no way depends on being able to reduce these sciences to a molecular level. At the very least, biology has autonomy as a descriptive science.

Descriptive autonomy, of course, is not the same as theoretical autonomy. The fact that biology employs numerous terms that are distinctively biological, in the sense that their primary application is to biological subject matter alone—terms like 'chromosome,' 'embryo,' 'hormone,' and 'mitochondrion'—does not establish biology as an autonomous discipline. Theoretical autonomy requires that these concepts be incorporated into a set of propositions which comprise a system of laws. If biology is to count as an autonomous science in this sense, its propositions must have the status of laws of nature.

On first reflection it might appear that any relation asserted between biological concepts that is empirically true should qualify as a law, at least at a low level. Smart has maintained (*135*, p. 58), however, that biological propositions are merely generalizations and that there are no biological laws in any strict sense. Not even the principles of Mendelian inheritance have the status of laws, on this view, because of their susceptibility to falsification. We do not have laws and theories in biology, Smart believes, for the same reason that we do not have them in electronics or chemical engineering: we are dealing with complicated systems which may very well be constructed in accordance with different principles at another place or at another time.

We have seen, however, that the susceptibility of biological laws to unanticipated falsification under circumstances which turn out to be beyond the scope of the law does not make them logically dissimilar to the laws of physics and chemistry. A distinction may indeed be drawn, as Nagel has done (*99*, chap. 5), between experimental laws and theoretical laws, where only the former express relations between observable (or experimentally determinable) traits. Even if we accept the distinction (which Nagel acknowledges is a vague one), however, it will not support the distinction Smart wants to make between physical laws and empirical generalizations. In the first place, given the theoretical nature of the concept of the gene, for example (see chapter 3 below), the laws of genetics would come out on the "wrong side" of the distinction. Secondly, experimental laws in physics, such as Ohm's law or Boyle's law, turn out to be very much like the ones in biology. None of these attempts to demonstrate a difference in kind between laws of nature and empirical generalizations of biology seems to have been successful.

Biological propositions may be looked upon as descriptive generalizations, as laws of nature and theoretical principles similar to those of physics, or as something in between. The first of these positions interprets biology as an unsystematic set of applications of chemistry and physics; the second treats biology as quite systematic and as having its own fundamental principles and theoretical structure; and the third may view the set of biological statements as quasi-systematic and as comprising what amounts to a distinct but rather loose science in its own right. According to the first, biology has neither laws nor theories; according to the second, it has both; and according to the third, it has laws but no theories. (No empirical science could be said to have theories but no laws, for such theories would lack empirical content.) Since, as I have argued, the criterion of universality provides no adequate basis for distinguishing the physical from the biological sciences, there are no grounds for denying the status of laws to at least some of biology's propositions. It does not follow from the fact that much of modern biological research is concerned with the application of chemistry and physics to biological regularities that the biological propositions that are thereby explained cannot themselves be laws of nature, embedded in a system of biological laws and theories.

Those who have sought to defend the autonomy of biology have generally gone beyond the claim that biology has its own laws and concepts, however. Biological autonomy has often been taken to mean that biological processes should be explainable in terms of fundamental principles which are themselves biological principles, and not laws of chemistry and physics. The so-called "organismic biologists," for example, such as E. S. Russell (*116*), J. S. Haldane (*52, 53*), and Ludwig von Bertalanffy (*12, 13*), have repeatedly urged that the temptation to search always for mechanistic or physicochemical explanations of organic phenomena be resisted, and that biology be recognized as a science with a logical and conceptual structure all its own. Thus von Bertalanffy believes that biology would have developed better as an independent science had it not been perturbed by neighboring sciences: "Growing up under the shadow of physics it has languished like a plant deprived of light" (*12*, p. 189).

A fully systematic and autonomous science of biology such as the organismic biologists envision would have to exhibit its own first principles, from which the laws pertaining to the individual fields within biology could be deduced. According to von Bertalanffy, these principles would be system-laws, fundamental principles of hierarchical organization. The mode of organization within biological systems, as J. H. Woodger has pointed out (*167*, p. 277), may embody relations that, though they may occur universally among biological entities, are anything but simple and do not exist at all among objects lacking this type of organization, that is, objects of the inanimate world. Fundamental principles suggested by von Bertalanffy (*12*, pp. 124-25) from which these characteristically biological relations would be deducible include a law of biological maintenance or organic equilibrium and a principle of hierarchical order, both static and dynamic. The task of biology, according to this program, is that of establishing the system-laws at all levels of the living world.

Such laws, if they could be formulated and shown to be universally true of living systems, would not represent an application of the laws of chemistry and physics, for they would be laws on a higher level. Each level of organization could be expected to have its own set of fundamental principles. What we would have would be the study of organisms of various sizes; biology, as Whitehead put it (*160*, p. 103), would simply be the study of the larger organisms, whereas physics would be the study of the smaller organisms.

It is worth pointing out that in viewing biological systems in this manner von Bertalanffy does not and cannot consider them as unique in principle. An organism, for him, is "a hierarchical order of open systems that maintains itself in the exchange of components with its environment by virtue of its system conditions" (*13*, p. 129). It thus turns out that the branch of science that contains these principles is not biology itself but rather what von Bertalanffy has named general system theory. But the important point is that it will be from these principles, and not from the laws of chemistry and physics, that the principles of biological self-regulation and organic growth are to be derived, if they are to be derived at all.

The ultimate test of a science's systematic nature and logical autonomy is axiomatization. Strictly speaking, if biology is to stand by itself, it must constitute a formalizable system, and this system will be logically independent of concepts and laws of physics and chemistry. The only portion of biology that has thus far been completely axiomatized is classical genetics, which J. H. Woodger has done (*166*) using the formal apparatus of Whitehead and Russell's *Principia Mathematica*. By translating the concepts and principles of genetics into this logical notation, Woodger was able to show how the predicted consequences of Mendelian pairings can be deduced from the science's basic axioms. To axiomatize biology as a whole, however, would be quite a different matter, since it would require a deduction of the first principles of all of the various biological subdivisions as well as the theorems that fall within each of these domains. What this would amount to would be the axiomatization of a portion of general system theory.

The point of attempting to axiomatize biology, I have suggested, is to assess the extent of its autonomy, as well as to determine its logical structure as a body of knowledge. For someone like Smart, who considers biological propositions to lack the universality of applicability that axiomatization seems to presuppose, the entire enterprise of formalizing biology appears radically misguided, analogous to trying to axiomatize the principles of electronics or bridge-building (*135*, p. 52). Smart is certainly correct in suggesting that it would hardly make sense to attempt to axiomatize a discipline unless we at least *believed* that the axioms themselves correspond to true first principles of a science. But it is precisely the extent to which biological principles are believed to have at least a provisional universality that provides whatever interest there is in axiomatizing biology or any of its branches: they apply to all living things. The program of axiomatization can claim cogency according as biological objects acquit themselves in conformity with the principles of the system. As long as living things are all generally similar to one another with respect to the features represented in the axioms, and dissimilar in these respects to anything that is not a biological object, axiomatization may be sensible. It is important, therefore, so far as axiomatization is concerned, that the earth, at least, contain no counterexamples. And as for the point that biology as thus axiomatized is probably a purely terrestrial science, this apparent implicit restriction need not constitute a significant objection, since, as I have argued, no scientific theory can be assured to have absolute universality.

The axiomatizability of biology is a precondition both of its theoretical autonomy and of its reducibility to a more fundamental science. The controversy over reduction in biology, which can be seen as an up-to-date version of the traditional mechanism-vitalism dispute, is essentially over whether an organism is "nothing but" an arrangement of chemical substances organized and interacting according to the same principles as apply to inanimate matter, or whether it is in some sense an entity "over and above" the aggregates of matter of which it is composed. A reductionist is one who maintains, and an antireductionist is one who denies, that physics and chemistry can ultimately explain all of biology.*

According to what is ordinarily taken to constitute reduction, for one theory to be reducible to another theory is for its propositions to be deducible from those of the more "basic" theory together with appropriate definitions and connecting principles. A paradigm case of this is the reduction of classical thermodynamics to statistical mechanics. Biology, at least presently, does not qualify for reduction in this sense, if only for the reason that it has not been sufficiently axiomatized to permit a formal logical derivation of its propositions.

Even if biology were completely axiomatized, however, reducibility might still not be guaranteed. If reduction requires the establishment of a one-to-one correspondence between the primitive terms of the reduced

*For more detailed discussion of the question of biology's reducibility to physical science, see Schaffner *117, 119, 120;* Scriven *122;* Rashevsky *111;* Shapere *125;* Hull *65;* Roll-Hansen *113;* also Polanyi *106, 108,* and Causey *21.* On the logic of reduction in general, see Nagel *99,* chap. 11; Kemeny and Oppenheim *72;* Feyerabend *36;* Schaffner *118.*

theory and terms of the reducing theory, lack of extensional equivalence between these terms may prevent reduction from occurring. Biological entities are picked
out on the basis of problems on the macroscopic level,
and these need not be assumed to correspond directly
with the objects in terms of which descriptions on the
microlevel are framed. There is no discrete biochemical
entity that corresponds to the gene for albinism, for
example, but rather a certain DNA sequence which acts
in combination with other factors to produce the characteristic functionally so designated. What determines
the entities and processes to which biological terms refer
may be a sequence of historical events which are individ-
uated according to biological rather than chemical concepts, and there is no reason to expect that chemical
and biological concepts used to designate a given structure will possess the same limits. What the molecular
biologist takes as given—namely, the particular modes of
organization that he undertakes to describe in chemical
terms—has a biological characterization in terms whose
epistemological basis has no necessary connection to
anything on the microlevel.

In order for there to be a successful reduction of biology to chemistry, analogous to the reduction of thermodynamics to statistical mechanics, there would have to
be a reformulation (or "correction") of its expressions
so that extensional equivalences can be established. Biological terms, such as 'gene,' would have to be redefined
and made much more precise if they are to become
applicable to the referents of chemical expressions. Furthermore, as Shapere has pointed out (*125*), any formal
reduction would have to take into account the idealiza-
tions which characterize physics and chemistry and
would require an analysis of the techniques of idealiza-

tion, approximation, and simplification. Schaffner (*119*), while noting these difficulties, has argued that this program of converting a theory into a form that can be reduced (by the construction of what he calls "reduction functions" for entity and predicate terms) is not only possible but extremely desirable: reduction can give new information about the domain of the reduced science, and can increase its predictive power while at the same time helping us to know the limits of its propositions.

If Schaffner is right, then biology is ultimately reducible to physics and chemistry, in the sense that all the phenomena designated in biological terms are explainable in terms of the laws of physics and chemistry. A view that represents a radical dissent from this position is that of Michael Polanyi, who has argued that biology is irreducible *in principle* to chemistry and physics. According to Polanyi, the fundamental mistake of those biologists who have accepted the reductionist position is that "they assume that a machine based on the laws of physics is explicable by the laws of physics" (*108,* p. 38). The reason he offers for believing that this *is* a mistake is that the structure of any machine is such that it cannot be derived from the laws of inanimate nature which govern those processes that are "harnessed" within the system as a whole (*106*). This structure serves as a boundary condition and is extraneous to the process which it delimits. Mechanisms of any sort, whether man-made or biological, are "boundary conditions harnessing the laws of inanimate nature," and are themselves irreducible to those laws.

Polanyi's argument rests on the fact that it is not possible to deduce the specific organization of the hormonal-enzymatic system or a particular sequence of bases in a DNA molecule from laws governing particulars on the molecular level, just as it is not possible to derive a vocabulary from phonetics or a grammar from a vocabulary. In each case the morphology of the higher level of organization acts as a boundary condition which "harnesses the principles of a lower level in the service of a new, higher level" (*106*, p. 1311). Organisms are held to be irreducible to the laws of inanimate nature for the same reason that clocks are: they are defined by operational principles that are embodied in matter by artificial shaping. The conclusion he draws is that, because these principles lie outside of physics and chemistry, the operations which they govern are not explicable in terms of these sciences.

Many, if not most, scientists would regard the conclusion that the operation of a clock cannot be accounted for in terms of physics as manifestly absurd, but Polanyi does not shrink from it. His argument is sound, provided one accepts the inference from the nonderivability of higher-level operations to their irreducibility and inexplicability in terms of the laws governing its particulars on the next-lower level. We cannot, in fact, deduce from the laws of physics the particular configuration of solids that make up a clock or any other mechanism, nor can we deduce the particular processes by virtue of which it is a clock or other mechanism. We may also concede that these operations are irreducible, if we make as a requirement for reducibility that the initial conditions—that is, the arrangement of the system's parts—be derivable from physicochemical theory, rather than taken as given. But to argue from irreducibility in this sense to the conclusion that "the operations of a higher level cannot be accounted for by the laws governing its particulars forming the lower level" (*108*, p. 36) and that "a mechanically functioning part of life is not explicable in terms of physics and chemistry" (p. 42), is to place what would seem to be an arbitrary restriction on what is to count as explication. It is in effect to disallow all explanations of laws and processes in terms of the behavior of entities on a microlevel. Given the ordinary uses of the terms, it is simply false to suggest that laws on higher levels are not sometimes explained in terms of laws on lower levels, or that complex processes are not sometimes explicated in terms of the physics and chemistry of processes occurring on the microlevel.

A crucial feature of any mechanism that distinguishes it from other configurations of matter is that it represents only one of a very large number of possible arrangements, all of which are equally compatible with the laws of physics and chemistry. It is for this reason that it is not possible to deduce the particular structure of a machine or organ from these principles alone. Higher levels of organization *are* determined by lower levels, in the sense that they must be compatible with the laws governing all processes on those levels, but they are not *uniquely* determined.*

This seems to be what Polanyi has in mind when he speaks of "the transcendence of atomism by mechanism" and the "harnessing" of the laws of inanimate nature by the structure of a machine. It may be that Polanyi gets carried away by his own modes of expression. Thus when he says that "the morphology of living things transcends the laws of physics and chemistry" (*106*, p. 1310), he seems to be suggesting that there is a total lack of connection between higher-level mecha-

*Essentially the same point is made by Causey (*21*), when he distinguishes between an explanation of the *existence* of a particular structure and an explanation of its *empirical possibility*.

nism and lower-level laws. This is obviously false, as Polanyi recognizes. Had he said simply that the existing morphology is but one of a large number of possible ones consistent with current theories of chemistry and physics, no such misimpression could have been engendered.

Similarly, Polanyi's use of the harnessing metaphor, with its implications of dominance and control, suggests that the laws of physics and chemistry have a status that is essentially subordinate to the principles that govern high-level processes. What actually seems to have happened is a shift from speaking of physicochemical *processes* being harnessed by mechanisms within which they occur to speaking of lower-level *principles* being harnessed by the boundary conditions imposed by the structure of a machine. We do in fact construct, and sometimes discover, systems in which physicochemical processes occur and can be controlled, and it is not extraordinary to describe such systems as "harnessing" these processes, as in the case of "harnessing" thermonuclear reactions within a reactor. Nothing "harnesses" the laws according to which these processes occur, however, either literally or metaphorically. The boundary conditions of which Polanyi speaks are not boundary conditions of the laws of physics or chemistry (in the way that extremes of temperature and pressure constitute boundary conditions for the Ideal Gas law), but they are boundary conditions for the occurrence of the internal processes. Physics and chemistry explain what they explain wherever it occurs. What is insidious about the assertion that it is the laws themselves that are harnessed is that it suggests that they could not possibly be expected even to explain how the machine that harnesses them works, not to mention how it came into existence.

It is important to realize that a machine or other physical system may have been constructed according to
principles that need not have been invoked in order to
account for the system's internal functioning. What this
shows, however, is not that such a system is inexplicable
in terms of physics and chemistry, but only that an
account of how a system works is not the same thing as
an account of how it came into existence. It may very
well be possible at some point to render an account of
the history of the system in terms of an expanded physicochemical theory, but the satisfaction of such a requirement is not relevant to the task of explaining how
a given system works. Even if biological organisms are
ultimately to be understood as the products of divine
creation, this would not detract from the adequacy of
an account of how these products work, any more than
a biblical account of the creation of the universe de-
tracts from the adequacy of a Newtonian account of the
operation of the solar system.

There is, nevertheless, a consideration that suggests
that the problem of biology's reducibility may not be so
simple as often supposed, especially by molecular biologists. This is the problem of accounting for the "hereditary" property of living matter, the exceptional relia-
bility of hereditary transmission in living systems. Although much has been achieved in attempting to work
out the mechanism of the self-replication of genetic material, there have not been even any loose treatments of
the physics of the processes which allow these structures
to persist in a relatively disordered environment. The
point is that this problem, unlike the other one, is neither simple nor understandable by classical models. If
the processes of inheritance are to be expressed in the

language of physics at all, it seems likely that it will have to be in the language of quantum mechanics.* Biology, even within the areas in which the molecular approach has produced the most striking successes, is still a very long way from having been shown to be reducible to physics and chemistry.

At this point it may be appropriate to consider the importance of the entire enterprise of attempting to reduce biology to physics and chemistry. What is the significance of being able to show that "in principle" one could deduce all biological phenomena from the laws of physical science? Perhaps it would be sufficient to show that these phenomena are *compatible* with physicochemical principles. After all, microstructure can be quite irrelevant to descriptions and explanations on the biological level. Whether or not one actually can deduce biological facts may be uninteresting, for the same reason that the nondeducibility of economics from physical science, unsupplemented by psychological principles such as might be needed to account for the effect of upbringing on economic behavior, for example, is uninteresting. If reduction means deduction, it may require appeal to principles of environmental determination which would go far beyond the domain of interest with respect both to the reduced science and the reducing science.

If the reductive approach to biology has any value, it must be that it can provide new insight into the behavior of entities within its domain. It has also been argued that explication of biological phenomena in physicochemical terms can serve as a means of rendering biological categories more precise, and thus to make more

*For a discussion of this point, see H. H. Pattee, in *155*, 2:279ff.

explicit the relations between biological knowledge and other areas of scientific knowledge. On the other hand, as a number of people have warned (see, e.g., *3*, *24*), the pursuit of reduction, if it becomes the central concern of biologists, can have the effect of prejudicing the subsequent development of theory, since it may inhibit the extension of traditional methods of inquiry such as have always been necessary for locating areas of interest. Biology as a subject exists as a result of having taken certain types of systems as given, and its concepts have developed in response to demands imposed by the environment on a macroscopic level. If reducibility is taken to imply deducibility and thus suggests the feasibility of rendering the reduced science and the reducing science logically isomorphic, it may be more than an innocuous myth: it could be a pernicious one.

If biology were sufficiently systematic to be capable of axiomatization but still could not be logically derived from propositions of chemistry and physics, it would obviously be called autonomous, regardless of whether it was considered compatible with these other disciplines. If, on the other hand, as seems to be the case, biology lacks this systematic character and yet includes within it a number of areas of research in which findings are rendered in exclusively biological terms and knowledge of microstructure is treated as irrelevant, it can be said to enjoy a kind of practical autonomy. The issue over whether a particular branch of science can be considered autonomous in this sense is a pragmatic one, in that a satisfactory answer may depend on how useful the science in question is when carried out in ignorance of its position with respect to the more "basic" sciences. And this, in turn, will depend upon the stage of development of both sciences. Astronomy today seems not to be able to achieve much explanatory success without physics, and hence has little claim to autonomy. Geology, on the other hand, appears to be carried out somewhat more independently. In this case, one can call it autonomous without seriously doubting its relation to physics; the reason here seems to have to do with the inanimate nature of its subject matter. Biology, which has had a long history of being carried out quite independently of physics, is perhaps the paradigm of a functionally autonomous science.

It has become a commonplace to point out that biology is not a theoretical discipline. It has theories, to be sure—theories of inheritance, theories of neural functioning, theories of gene replication—but it exhibits no hierarchical structure of laws and theoretical principles. There is no theoretical biology in the way that there is theoretical physics, for it is only within various isolated areas of biology that biological propositions can be arranged in systematic order. Thus Woodger has ridiculed the notion that Darwin is the Newton of biology: "To suppose Darwin a Newton is to suppose biology to have reached a degree of theoretical development comparable with that of physics in the eighteenth century, which is preposterous" (*167,* p. 483). Biology, in Woodger's estimation, has yet to find its Galileo.

Woodger may have gone too far in disparaging the achievements of his science, however. It is indeed true that biology can claim nothing even remotely resembling the achievement of a Newton. But it may perhaps fruitfully be asked what it would mean for any science, including physics, to have "found its Galileo." If there is a Galileo of chemistry, it is presumably Lavoisier, whose contribution was a revolutionary new classification of substances that provided an effective and systematic means of understanding a large variety of chemical changes. Essentially the same sort of revolution had been effected by Galileo himself, who, by focusing on a new central notion (rectilinear motion at constant velocity) and elaborating its consequences, produced a new way of classifying the motions of bodies. It is at least worth arguing that the principal significance of Galileo's achievement is not that he employed a deductive mathematical formulation, but that he gave a central role to certain concepts which, when developed mathematically, yielded a systematic treatment of a class of phenomena that previously had submitted to no such account.

When we observe, moreover, that in biology the Darwinian theory of evolution by natural selection provides the basis for modern taxonomy, as well as the conceptual background for explanations in anatomy, ecology, physiology, and ethology, we are strongly tempted to consider Darwin to be in a very significant sense the Galileo of biology. The elaboration of the theory has not been mathematical or rigorously deductive, to be sure (except in a few restricted areas), but neither has the elaboration of modern chemistry. If a "Galileo" is thought of primarily as someone who contributes a unifying conceptual scheme that creates the *possibility* of developing a systematic theoretical framework, then biology may indeed have had its Galileo, if not its Newton.*

As far as formal theoretical structure is concerned,

---

*An interesting alternative conception has been suggested by Manser (*84*), who prefers to see Darwin as biology's Karl Marx, having provided "a basic picture which seems to render a complex mass of facts comprehensible without giving either the power of control or of prediction" (p. 30).

physics is more highly developed than chemistry, which is more highly developed than biology. The reason for this progression appears to reside in the increasing complexity of the subject matter; a science that deals with objects a greater variety of whose features are considered relevant to that science will be less susceptible to being cast in axiomatic form, or shown to have a formal systematic unity, than one that concerns a relatively simple set of properties. But a science which, like biology, does not yet contain sufficient higher-level principles to provide covering laws under which the phenomena in its domain can all be subsumed and hence explained may have available to it explanations of a different sort: these phenomena may submit to explanations in terms of the principles of the more basic and better systematized sciences, provided their objects can be assumed to comprise the ingredients of the particular aggregates with which the science in question deals. As the formal standard, the objective of systematic unity within a science, becomes increasingly difficult to satisfy, it may be less and less important that it ever be satisfied, for the availability of the alternative approach provides the possibility of explanation on another level. Biology has and needs two directions for its investigations: upward toward system-laws, and downward toward chemistry and physics. The science of biology can be considered at a primitive stage of development only if the second of these directions is ignored. It will deserve to be considered mature, however, only when it can exhibit principles on all of the levels at which the matter of life is organized.

2. Biological Explanation

An explanation is a statement or set of statements that confers intelligibility upon an event or state of affairs which is thereby said to be explained. An explanation is typically the answer to a "Why" question, where such a question designates something that the asker wishes to understand. Scientific explanation is what brings about scientific understanding. To comprehend the nature of scientific explanation is thus to know what counts as scientific understanding, to know what counts as a scientifically adequate answer to a corresponding "Why" question. Our present concern is with the kind of understanding afforded by explanations in biology.

Since explanatory adequacy depends upon understanding and understanding presupposes a cognizant individual, the problem of characterizing explanation must have a pragmatic as well as a purely formal or logical aspect. An explanation must work; it must satis-

fy a need. For any explanation to be effectively explanatory, it must be presented against a background of recognized phenomena for which it is assumed that no
explanations are necessary. That which is to be explained must be explained in terms of things that do not
need to be explained, things that the recipient of the
explanation either believes he already understands or
comes to accept as needing no explanation beyond the
account given him. Thus the adequacy of an explanation
of why moisture forms on the outside of a glass of cold
water will depend, if not on a familiarity with the general phenomenon of condensation of vapors upon cooling,
at least on coming to accept the principles offered in
explanation as terminal points of inquiry. Explanations
come to an end somewhere, Wittgenstein reminds us.
Characterizing biological explanation consists both in
uncovering the logical structure of explanations of this
sort and in determining what kind of bedrock they may
be presumed to rest upon.

The fundamental question that any proffered explanation purports to answer is: How does it happen that X is
the case (rather than Y)? According to what has come
to be regarded as the "classic" approach to the logic of
explanation, a scientifically adequate answer to any
such question will consist of showing how a statement
asserting the occurrence of X, the event to be explained,
is deducible from general laws in conjunction with statements of antecedent circumstances. This account, attributable to Hempel and Oppenheim (*60*), is known as the
deductive-nomological or covering-law model of scientific explanation, and is also called the predictive account, since it requires that explanation and prediction
exhibit the same formal structure, differing only with
respect to the temporal relations between the event ex-

plained or predicted and the statements offered as either explanation or prediction.

An alternative conception of scientific explanation, suggested by Toulmin (*148*) and Kuhn (*75*), is one that views explanation (at least in some cases) as a means of exhibiting a phenomenon under investigation as either a special case or a complex combination of certain previously recognized intelligible types. According to this account, explanation consists in relating a particular happening to something else that we regard as needing no explanation. As Toulmin expresses it, scientific explanation consists in "relating the anomalous to the accepted" (p. 61). Scientific understanding, according to this viewpoint, resides not in determining how an event could have been predicted or deduced from a knowledge of general laws and relevant antecedent circumstances, but rather in being able to represent it in terms of certain paradigms or explanatory ideals that we take to be already understood.

These two approaches to the nature of scientific explanation may not be entirely incompatible, since they can be viewed as representing attempts to answer somewhat different questions. Whereas the former approach is aimed at discovering the logical relationships between the statements used to describe phenomena and the statements that serve to explain them, the latter is concerned with indicating what apart from its logical structure determines the explanatory adequacy of a scientific explanation. There will be collisions between these two approaches, to be sure, particularly over the role of prediction and predictability, but it will still be useful to explore them separately, and to inquire of biological explanations both as to their logical structure and as to their pragmatic standards of acceptability. Such a program calls for an investigation of what various instances of biological explanation have in common, both with respect to the types of statements they involve and with respect to the nature of the unexplained or self-explanatory paradigms upon which they can be shown to rest. In the present chapter I shall concentrate on the former aspect of this investigation; I shall defer consideration of what paradigms are exploited in biology until chapter 4, in which the basic underlying conceptual systems upon which biological science rests will be scrutinized.

As a means of immersing ourselves in some of the problems of biological explanation, consider the following list of statements:

1. Dinosaurs are extinct.
2. The shark's intestine contains a spiral valve.
3. Some humans have harelips.
4. Leaf-dwelling insects are green, whereas bark-dwelling ones are gray.
5. A blue-eyed child may be born to brown-eyed parents.
6. The incidence of lung cancer in the United States has more than doubled in the past fifty years.
7. Cats see better at night than humans do.
8. Green plants turn toward the sun.
9. Among some kinds of birds, only the male of the species is brightly colored.
10. Mules are sterile.
11. Many species of animals maintain constant body temperature under varying external conditions.
12. Some species of animals carry their young in pouches.

Each of these statements calls attention to a biological fact to be explained. If we consider the sorts of explanations they might be expected to elicit from biologists, we discover a significant variety in types of account. Facts (7) and (11), for example, seem to call for explanations of a causal nature, in terms of general laws and antecedent or concomitant circumstances. Facts (1), (5), and (12), on the other hand, would appear to demand accounts in terms of historically prior conditions, and (2) and (9) seem to suggest explanation in terms of purpose or function. Some of these examples may admit of more than one type of explanation. In any case, it seems clear that there is no single type of characteristically biological explanation that is appropriate for all biological phenomena.

Provisionally, we may say that there are prima facie three types of biological explanation, three different ways in which intelligibility may be conferred on biological phenomena. I shall call these types of explanation, respectively, causal, historical, and functional. A causal explanation is one that exhibits the deductive structure indicated by the covering-law model for scientific explanation. An example would be the explanation of sunburn in terms of the biochemistry of skin pigment or the explanation of the flight of an insect in terms of aerodynamics. Explanation in these cases consists in subsuming the phenomenon to be explained under general regularities, which could have been used, in conjunction with statements of relevant antecedent circumstances, to predict the event in question before it actually occurred.

Among the most prevalent of causal explanations to be found in contemporary biology are those in which biological phenomena are explained by subsuming them under principles of chemistry and physics. Explanations of this sort, in which properly biological statements appear only in the role of indicating antecedent circumstances, are typical causal or deductive-nomological accounts, similar to those found in the physical sciences. According to the conception of biology as ultimately biochemistry and biophysics (an interpretation discussed in the last chapter), all biological explanations are of this type or aspire to it. In addition, however, many explanations used by biologists do not employ physicochemical principles as covering laws but still exemplify this model of explanation. The fact that a principle that is used to explain a phenomenon may be a general statement of biology, such as might concern hormones or molar environmental constituents, for example, does not necessarily affect the logic of the explanation. As long as biological explanations attempt to subsume phenomena to be explained under laws, whether physical and chemical laws or laws of biology, they can be expected to fit the covering-law model of scientific explanation.

Historical explanations, which may or may not constitute a subclass of causal explanations (this being a topic of the discussion which follows), represent attempts to show how certain states of affairs have the character they do by appealing to statements descriptive of the earlier states of the system under consideration. Explanations of skin markings as scars resulting from earlier wounds, of neurotic behavior as the outcome of previous environmental influences, of physical deformities in terms of embryological irregularities, and of anatomical features of an organism by natural selection operating on its ancestors, all are examples of historical explanation. An explanation of this sort purports to show the phenomenon to be explained as a natural result of a historical process by identifying its relevant temporal antecedents.

A functional explanation, finally, is an attempt to indicate why something is the case by reference to certain ends that it promotes. Typically, a functional explanation is an account of the contribution made by a given structure or process to the maintenance of certain characteristic activities that are essential to the welfare of the system in which it is found. Such an account might be offered to explain why humans have thymus glands or why cows lick salt. Whereas a causal explanation of X is supposed to answer the question, "What principles and circumstances determined X?" and a historical explanation is supposed to answer the question, "How did X come to be the way it is?" a functional explanation is supposed to answer the question, "What does X do?" To give a functional account of something is to suggest the purpose for which it might have been created, if it did not already exist.

The questions that need to be explored with respect to these three types of explanation in biology are, first, whether and in what sense each of them really does explain; second, how these types of account are related to each other; and third, whether any of them is uniquely peculiar to biology. The answers to these questions will have a direct bearing on the issue of biological science's relation to chemistry and physics, on the one hand, and descriptive natural history, on the other. The following discussion will concern explanations of the type in which appeal is made either to the past history of the organisms whose characteristics are to be explained or to their ancestry; functional accounts will be deferred for consideration in a later section of this chapter.

Historical Explanation

A number of positions are possible with respect to the ways explanations in biology of an apparently historical character are to be understood. The range of these interpretations is covered, I believe, by the following four possibilities:

1. Biological explanations may be historical only in the respect that the statements of antecedent circumstances include historical facts. An example of this type would be an account which mentioned a previous injury or an ancestral characteristic in explaining the presence of some structural anomaly. The only laws involved in such a case might simply be process laws, either on the level of chemistry or physics or on the level of biological structures.

2. Biological explanations include historical laws that are ultimately explainable in terms of nonhistorical laws. Examples of such laws are the laws of Mendelian inheritance and classical genetics. Laws of this type, which connect present phenomena with prior ones, might be invoked to explain the incidence of certain observable traits, but are themselves subject to explanation in terms of physicochemical principles.

3. Biological explanations may involve appeal to laws that are irreducibly historical. On this view, certain biological explanations might involve laws for which there could conceivably be no possible physicochemical explanation. Or it might be maintained that the possibility of explaining one of these laws in nonhistorical terms is

irrelevant to the essentially historical status of the explanation. Thus it might be claimed that molecular biology can never destroy the historical character of a phylogenetic or Mendelian explanation, and that neurophysiology can never supplant laws of association in psychology, even if it is assumed that all behavior and mental activity is itself ultimately explicable in terms of neurophysiological principles.

4. Biological explanations need not contain laws at all, and hence they are capable of falling outside the deductive-nomological or covering-law model of scientific explanation altogether. According to this conception, the explanation of phenomena such as the protective coloring of insects or the persistence of apparently useless structures (e.g. the vermiform appendix, or the wings of a nonflying bird or insect) need appeal only to statements of antecedent conditions and possibly some generalizations of natural history.

It is possible, of course, that all four of these conceptions may have genuine instances, and that the class of apparently historical explanations in biology may actually embrace several distinct types. Our concern at this juncture, however, is not so much with the classification of these explanations as it is with the question as to whether any of these apparently historical explanations in biology constitute counterexamples to the covering-law theory of explanation. Does biology provide any accounts which are explanatory but not causal?

Ontogenetic and phylogenetic explanations are the obvious candidates for noncausal or nondeductive historical explanation in biology: explanations either of the development of characters from earlier stages of the individual organism or of the evolution of characters from previous generations. Ontogenetic explanations, however, such as might be presented to account for certain morphological features in terms of the embryological development of a fertilized egg under given environmental conditions, are subject to characterization either as essentially causal or as not explanations at all. For if an explanation is sought for a certain juxtaposition of organs or area of pigmentation, an answer in terms of embryological structures and developmental processes might be challenged as being no more explanatory than would the attempt to "explain" metal fatigue in terms of aging or an epidemic in terms of the spreading of an infection. Such an account could be explanatory only to the extent that it eliminated the possibility that the phenomenon in question might have arisen from other prior conditions. On the other hand, if one of these accounts indicated the mechanism of a given developmental process, it would then amount to a straightforward causal account in terms of physicochemical principles. Embryological or ontogenetic explanations, whether on the level of molecular biology or that of subcellular biological particles, will exhibit the deductive-nomological form as long as they contain general principles.

Most of classical embryology, it should be noted, was carried on without much concern for general principles of any sort. Ontogenetic accounts typically showed how one structure develops into another one. The central nervous system of vertebrates, for example, develops from a tube formed as a result of the closing of an oval-shaped plate on the dorsal side of the embryo. Propositions such as these have never been considered candidates for laws of nature, and they are usually not deemed explanatory. Descriptions of growth processes may be interesting, but as biologists will commonly acknowledge, they do not constitute explanations.

Phylogenetic accounts, on the other hand, appear to be more problematic. Such accounts attempt to show how it has come about that the members of adult populations of organisms possess the characteristics they do, where the process to be described seems to be neither causal nor one of continuous development within a single entity. If they are genuine explanations, they are evolutionary explanations. Will an account of, say, the evolution of the giraffe count as explanatory, and if so, what sort of explanation is this?

It has been argued that phylogenetic explanation, as well as ontogenetic explanation, is either pseudo-explanation or else explanation in the deductive-nomological pattern. Beckner's position (*9,* chap. 5), for example, may be summarized as follows: for a property of an organism to be explained phylogenetically, it must be exhibited as the outcome of a continuous chain of events that are causally related. If phylogenetic change is understood as a change in adult type of a succession of organisms, however, then the requirement of continuity is violated, since it is obviously not the same adult that has undergone gradual modification in the course of evolution. Only if phylogenetic change is construed as a succession of individual ontogenies, wherein the reaching of sexual maturity and formation of germ cells of one organism is immediately followed by fertilization and initiation of the ontogeny of another, can a phylogenetic explanation be recognized as a genuinely historical explanation. Such an explanation, Beckner maintains, must employ a general theory of heredity and would consist of an analysis of the hereditary factors plus a causal explanation of the successive ontogenetic development in terms of these factors. It is Beckner's view that if any property can properly be said to be phylogenetically explained, its presence must be deducible either from laws of ontogenetic development coupled with statements tracing the pathways of sets of hereditary factors (e.g. genes) through a succession of individual organisms, or else from a set of statements indicating the properties of ancestral members of the phylogeny together with historical laws connecting properties of earlier members of the phylogeny with the property whose presence is supposed to be explained. In both cases the pattern of explanation is deductive-nomological.

Beckner appears to have paid a considerable price for holding that historical explanations in biology must fit the deductive-nomological pattern, however. By requiring a historical explanation to relate the event being explained to a prior event by a *continuous* chain of intermediary events, he effectively disallows the possibility that there be any such thing as a historical explanation, even in history. It is seriously to be doubted that an effective explanation of the defenestration of Prague in 1618 or the Japanese attack on Pearl Harbor in 1941 in terms of chains of prior historical events needs to indicate a thorough temporal contiguity between succeeding events mentioned in tracing a historical sequence. It is important only that the omitted intervening events between the selected beginner-event and the event to be explained not be deemed relevant to understanding the event in question. And as far as historical laws are concerned, it is by no means clear that there are any such laws, or that any historical explanations need appeal to them. For if there were any general historical laws of the type suggested by Beckner, namely, "If the members of the phylogeny of an organism O were characterized by F_1 at t_1, F_2 at t_2, . . . , F_n at t_n, then O will be characterized by P" (*9,* p. 103) (where F_1, F_2, . . . , F_n are properties of members of the phylogeny, P is the property to be explained, and no organism in the phylogeny with an F also has a P), we should expect phylogenetic explanations to be predictive. As Beckner himself acknowledges, however, such explanations are in general not predictive. Either there simply are no phylogenetic (or other historical) explanations in biology or Beckner's account is mistaken.

There is another point at which Beckner's analysis may be challenged, namely, his contention that phylogenetic explanation must employ a general theory of heredity. If Beckner's position is that such an explanation must *presuppose the possibility* of a theory of inheritance, then I believe he is correct. However, this contention is quite different from the claim that a phylogenetic explanation must *employ* an inheritance theory. It is important to note that Darwin, by his own admission, had no such theory to put forward. Someone might wish to argue that Darwin's disclaimer constitutes an acknowledgment that he could offer no *full* explanation of the facts of phylogeny, and that he thus recognized that his accomplishments fell short of providing explanations. It does not seem reasonable, however, to refuse to regard any of Darwin's deliverances as explanations. Would it not be extremely pedantic, if not wrongheaded, to deny that Darwin's attempts to account for the distribution and concentration of species in terms of migration and descent-with-modification, or the existence of rudimentary, atrophied, or aborted organs in terms of disuse and natural selection, are really explanatory? The Darwinian theory of evolution by random variation and natural selection quite obviously does explain a large class of phenomena, at least to the extent

that it succeeds in rendering intelligible a previously in-
coherent set of similarities and differences. The fact that
it did not, at the time of its original enunciation, include
any sort of mechanism whereby the facts of variation
and inheritance could be accounted for indicates not
that the theory lacks explanatory power, but that certain principles that the theory presupposes or includes
are susceptible to further explanation. A proffered explanation is rendered no less explanatory by pointing
out that some of the regularities it expresses are themselves in need of explanation. To demand of an explanation that it leave no explanatory gaps is to require every
explanation to be an ultimate explanation.

Beckner acknowledges the usefulness of the sort of
account that attempts to show an event as a natural
consequence of a series of historically prior events, but
he declines to call such an account an explanation, pre-
ferring to use the term 'genetic analysis.' According to a
somewhat broader conception of explanation, such an
account might be considered quite explanatory, since it
is obviously capable of contributing to an understanding
of the phenomenon being considered. Whether or not
these accounts are taken as explanations or merely as
analyses, however, I believe that Beckner is correct in
arguing that they do not conform to the deductive-
nomological pattern of explanation. The point of this
kind of analysis, he notes, is not to establish that certain
historical events are causally connected, but rather to
show an event to be explained as the outcome of a
sequence of events that are implicitly assumed to be
causally connected. The correctness of a historical analysis requires that the events specified in the chain *be*
causally connected, but the connection need not be
asserted in the explanation and may in fact not be

known or specifiable in a theory. Whereas a causal or deductive-nomological explanation demands that there be and be known a set of general principles which would show that the occurrence of the earlier event, together with other factors, is sufficient for the occurrence of the later one, a strictly historical analysis requires no such knowledge. Causal laws are not part of historical explanation.

In explaining an event by citing a series of precursors, we often neither provide nor ask for laws or general principles. What sorts of laws could we expect to uncover concerning the connections between the members of a sequence of events the mere listing of which serves to make the event under consideration intelligible? Scriven (*124*) has suggested that these would be truisms, such as "if you knock a table hard enough it will cause an ink-bottle that is not too securely or distantly or specially situated to spill over the edge (if it has enough ink in it)" or "preventing attack is a good reason for invasion when victory is certain without too much fighting and moral considerations are not too highly regarded." They are not stated because they are not *worth* stating. They explain nothing because they exclude nothing. The function that one of these principles might be supposed to serve would be that of bridging the gap between two temporally separated events. However, the only way one can thus cross such a gap is by assuming that it can be crossed, that is, by begging the question. The way we can tighten up the connection between an object's being pushed and its falling is by requiring that it be pushed hard enough to fall. And where the connection between the events described is not sufficiently obvious to be immediately recognized as causal, the account fails to be explanatory at all.

As a further aid to distinguishing historical explanations from causal explanations in the deductive-nomological pattern, it is interesting to note the kind of response that is likely to be evoked by a charge of inadequacy directed to a historical explanation. At least one type of unsatisfactory historical explanation would be one in which the person to whom the explanation is given is justifiably unconvinced that the event to be explained follows as a matter of course from the earlier events cited. Someone may question the connection between a driver's turning around to attend to a crying child in the back seat and the subsequent overturning of the automobile, or the connection between the enactment of a new tax law and a subsequent insurrection. In each case the reply takes the form of an attempt to fill in the intermediate steps, subdividing the hiatus into a series of shorter intervals that can be run through without difficulty. A historical explanation thus purports to show the way from past events to a present one, not by appealing to a broader framework under which the present situation can be subsumed, but by pointing out particular landmarks by which the path is made apparent.

There is a type of historical explanation, then, that does not conform to the deductive-nomological pattern. The question is, does biology exhibit any such explanations? Ontogenetic accounts, we have seen, are explanatory only when they include specification of the chemical or subcellular mechanism of the changes described, in which case they fit the covering-law model of explanation. A phylogenetic explanation, on the other hand, such as might be given to account for the existence of the horns of the rhinoceros or the coloring of the monarch butterfly, may be explanatory without meeting the deductive-nomological standards. And these are not the only types of explanations within biology in which no specific causal mechanisms may be known and no general principles have been formulated: other examples are the explanation of a disease condition in terms of vitamin deficiency, and the explanation of a particular type of muscularity in terms of previous exercising. When considering cases such as these, one may be inclined to say, "There *must* be laws that govern these phenomena, or else we would never accept the explanation." There is indeed a certain force to this objection, despite its question-begging character. What is significant about such explanations, however, is that they may be put forward and found acceptable *before* any law is ever formulated. It is true that the adequacy of explanations of this sort depends on the existence of lawlike relations concerning the phenomena under consideration, but the fact that they often cannot be specified indicates that biology can indeed offer explanations that contain no general laws.

There is one further point to be made concerning the role of noncausal historical explanations in biology: it may be asked whether biology contains any cases in which the discovery of a causal mechanism is not to be considered a biologically relevant contribution. I believe that this question must be answered in the negative. If we consider the various cases of apparently noncausal historical explanations in biology, we find that there is always presupposed some sort of underlying principle that could, if it were known, be used to enable us to fill out the explanation by revealing the mechanism by which the changes described occur. In most cases, if not all, the types of mechanisms that biologists have presupposed in describing biological processes, whether ontogenetic, nutritional, or phylogenetic, have been physicochemical. But even the vitalists wanted to understand how particular vital processes are mediated. If embryological development and cell-differentiation are determined by entelechies, biologists are nevertheless interested in discovering the general principles that govern the way these things exert their influence. Biology, even when it is engaged in explicating individual and ostensibly unique phenomena, is concerned with discovering regularities and with identifying biologically relevant subprocesses, whatever their nature be. With respect to the efforts of biologists to understand how living organisms work, this concern is obvious. It is also present in their efforts to show how organisms and species got to be the way they are. Biologists are committed to seeking to explicate the mechanisms of change and to identifying the substrata in which biological changes occur. To believe that biological processes are not subject to such causal explication is to deny the possibility of a science of biology.

The contrast with explanations of events in the history of human affairs becomes especially striking at this point. History, or at least some history, is *not* fundamentally concerned with the discovery of regularities, but is rather concerned with rendering intelligible individual events. Furthermore, in history, when we attempt to explain an event by exhibiting it as the outcome of a sequence of prior events and states of affairs, we do not ordinarily ask for an account of *how* the events in the chain produce their successors. Historical explanations can be expected always to be given on the same level as the events to be explained, namely, that of human actions, social movements, and other molar phenomena. The connections between events that are presupposed in historical analyses, unlike those in historical explanations in the biological sciences, function as primitives for the purpose of conferring historical understanding. To attempt to go deeper in explicating the mechanisms of causation between historical events is to leave history as it is usually conceived and to enter the realm of psychology.

Returning to the four possible interpretations suggested earlier with respect to the nature and role of historical explanations in biology and their relation to explanations of the covering-law type, we see that all four positions can claim supporting instances. The explanation of heart failure as the result of arterial occlusion caused by chemical deposits originating in ingested foodstuffs, for example, illustrates the type of physicochemical explanation in the deductive-nomological pattern that includes historical facts among the relevant antecedent circumstances. Classical genetics provides explanations using historical statements as explanatory principles, which are themselves explicable in terms of molecular biology. At the same time, since no explanation need be assumed to be "ultimate," it can be argued that classical and other molar accounts can nonetheless be considered explanatory regardless of whether the historical principles they employ are known to be explicable in terms of nonhistorical laws such as those of physics and chemistry. Finally, we have seen that in the case of phylogenetic explanations, we can have genuine explanations in biology that involve reference to no general principles at all, provided we are willing to assume that the sequences of changes cited are subject to explication in terms of principles and mechanisms yet to be discovered.

Biology employs a variety of types of explanation, ranging from the strictly causal and mechanistic deduction to the plainly historical narration of sequences of events. Each may be adequate to its specific uses, in the sense that it is capable of rendering phenomena intelligible. But lurking behind every apparent noncausal explanation in biology is the specter of a possible causal one, even in the cases in which no relevant general laws are known. Biological explanations, including historical ones, stand in varying degrees of resemblance to a logical ideal or explanatory paradigm, namely, that which finds most common application in physics. Biology does indeed employ historical explanations of a noncausal sort, but it does not accept them as ultimate.

Explanation, Prediction, and the Theory of Evolution

One of the issues with which the problem of scientific explanation is most deeply entwined is that of the relation between explanation and prediction. An important feature of the covering-law model of scientific explanation, it was noted, is that the same formal analysis applies to scientific prediction as to explanation. Any event that can be explained by showing its occurrence to be deducible from statements descriptive of antecedent circumstances, in conjunction with general laws, could also have been predicted on the basis of the same deductive argument. Defenders of the covering-law theory of explanation have often used this notion of predictive potential as a criterion for distinguishing explanations from pseudo-explanations. Application of this criterion to explanations in biology, however, as well as in a number of other areas, appears to have the effect of disqualifying a number of accounts that might otherwise be regarded as explanatory. In particular, it might lead us to deny that the theory of evolution affords any explanation of the origins, survival, and distribution of species, since this theory apparently provides no basis for predicting the phenomena it purports to explain. In order to come to an understanding of how explanation and prediction in biology are related, therefore, it will be necessary to examine the theory of evolution in the light of some of the challenges which have been directed toward it.

The theory of evolution is a theory of the behavior of certain types of objects, namely, populations of organisms. It has sometimes been maintained that the theory possesses a deductive core, and that what Darwin was doing was showing that natural selection, divergence of character, and the extinction of less adaptive forms can all be deduced from certain empirical assumptions about populations and environments. A number of attempts have been made to indicate just what this deductive argument would be. Anthony Flew (*40*) for example, after citing Darwin's own disclosures that "the Struggle for Existence amongst all organic beings . . . inevitably follows from their high geometrical powers of increase (*29*, pp. 4-5)" that natural selection follows as a *consequence* of this struggle (p. 490), and that his "whole volume is one long argument" (p. 459), tries to reconstruct Darwin's argument and to show that natural selection is logically entailed by inheritable variation plus a struggle for existence, which is entailed by the conjunction of a geometrical rate of increase of organisms and the limitation of resources needed for their survival. The deduction he sees as one whose steps are short and simple and whose premisses are obvious facts.

Flew's argument, however, as Hull has pointed out (*65*), is not logically valid; all that one can deduce from the premises Flew presents is that not all organisms that are born will survive. In order for a struggle for existence to follow logically from a high rate of increase and limited resources, either "Struggle for Existence" will have to be defined in such a way as to include any form of differential survival whatever, or a statement will have to be added to the premises to the effect that when not all organisms survive it is a struggle that determines which ones do (ruling out the possibility, for example, that some will voluntarily starve). Similarly, if natural selection is to be deducible from a struggle for existence and inheritable variation, either the premises will have to specify that it is such variation (and not geological and meteorological catastrophes, for example) that determines differential survival, or whatever happens to survive will have to be considered, by definition, to have been selected for. In other words, in order for natural selection to be deduced, it is necessary to assume that all population changes can be seen as consequences of natural selection and must be produced by that mechanism and that mechanism alone.

It is the alleged deducibility of natural selection that has given rise to the oft-repeated charge that the theory of evolution is ultimately circular and hence empty of content.* Specifically, it has been objected that the notion of the survival of the fittest is itself tautologous, since fitness is defined only in terms of actual survival. Indeed, the validity of the deductive argument does depend on identifying surviving to reproduce with being the fittest, as Flew correctly points out (*40*, p. 5). What this shows, however, is not that the theory of evolution has no content, but rather that it is not a deduction. A formal deductive theory is always circular, in the sense that certain of its key notions must be implicitly de-

*See Manser *84;* Bunge *17;* Eden, in *91*, pp. 5ff. In rebuttal, see Goudge *47*, pp. 117-18; Flew *38;* Hull *65.*

fined in terms of others. What is logically contingent in a deductive theory cannot be the relation between the premises and the conclusions but rather the applicability of these propositions to the world. Biologists do not in fact determine fitness only on the basis of actual survival. What the principle of the survival of the fittest asserts in its empirical application is that the likelihood of an organism's or species' survival is dependent upon inheritable variations which affect its viability in particular environments. It excludes, among other things, the possibility that individual variations should be irrelevant or insignificant to the likelihood of organisms having fertile descendants.

Darwin's argument in the *Origin of Species* was not so much an attempt to deduce what *should* occur as a result of reproduction with variation, limitation of resources, and so on, as an effort to show that what does happen can be seen as direct causal consequences of these states of affairs. That he recognized that *actual* survival cannot be deduced from his empirical premises is suggested by the fact that he hedged his argument by identifying being naturally selected with having a better *chance* of surviving by virtue of inheritable variations (*29,* p. 5). In point of fact, Darwin did not make a clear distinction between what can be expected to occur and what actually does occur. Had he done so, he might have seen that what could be deduced—namely, which organisms have the greatest likelihood of surviving—may very well not coincide with what actually happens, and that what actually happens—the survival of those organisms that do in fact survive—could not be deduced on the basis of his theory.

The content of the Darwinian or neo-Darwinian theory of evolution is that natural selection occurs and can be used to account for, among other things, the specific similarities and dissimilarities among organisms past and present without invoking such "unnatural" occurrences as catastrophes and special acts of creation, or such empirically unverified types of process as inheritance of acquired characteristics. As explanations, however, these accounts do not fit the deductive-nomological pattern, owing to the fact that we lack all of the relevant general principles by virtue of which individual variations and environmental conditions and their efforts on types of organisms might be deduced. Lehman has suggested that evolutionary explanations can be made to fit the statistical-probabilistic variant of the covering-law model, according to which that which is to be explained need not be implied with deductive certainty but only with high probability, and he has attempted to show how various forms of adaptation can be explained in this manner (79). A number of examples of evolutionary explanation, such as that of industrial melanism in moths, can be accommodated by this account, but when it comes to phylogenetic explanations in which a given character or set of characters might be represented as the culmination of a long sequence of changes, it is difficult to see how any sort of covering-law model of explanation could fit. There seems to be a definite disparity between the requirements of the covering-law model and our ordinary standards of explanatory adequacy, at least as far as explanations in evolutionary biology are concerned.

Given that the theory of evolution nevertheless provides the basis for explanations of a sort, regardless of how poorly they may be found to conform to certain logical forms, one may ask about the predictive capacities of the theory. Scriven, attempting to show that evolution is a theory that explains but does not predict, has argued that the most Darwin's theory is able to provide are hypothetical probability predictions, such as the proposition that organisms that happen to have the ability to swim would be likely to survive if a flood occurred (*121*). In addition, it should be noted that since mutations are random and hence unpredictable, there are cases of evolutionary explanation in which even hypothetical probability prediction is impossible. We can explain an organism's adaptation to an altered environment in terms of the natural selection of a series of mutants, but even if we had full knowledge of the environmental conditions we could not predict what sort of mutants would be produced. The number of possible ways organisms could become adapted to a given environment far exceeds the number of actual adaptations and is indeterminate. Hence the problem of predicting which organisms will survive is not simply a matter of our inability to predict environmental changes. The theory of evolution affords explanations of phylogenetic changes not by showing how they *had* to happen, but only showing how they *did* happen.

The theory of evolution enables us to explain what we could not have predicted because we know what it is that we wish to explain. We could not reasonably expect to have been able to *predict* the evolution of the horse from a four-toed to a three-toed and then to a one-toed animal, or the evolution of black-winged moths in industrial England, and yet we are able to explain these phenomena in terms of the natural selection of mutants. Scriven has suggested that the reason that such evolutionary explanations do not apparently afford potential predictions is that the only evidence we have for asserting some of the statements essential to the explanation is that the event to be explained in fact occurred (*121*). Analogous cases that he cites are the explanation of the collapse of a bridge on the hypothesis of metal fatigue intense enough to cause failure, and the explanation of a man's murder of his wife as the result of intense jealousy (*123*). When constituent hypotheses of an explanatory account are supported only by the occurrence of the event to be explained, it is clear that no explanatory argument could have been used to predict that event.

In defense of the covering-law model and the symmetry between explanation and prediction, Hempel has replied that Scriven's argument by no means shows that it is either logically or causally impossible for us to have foreknowledge of the relevant explanatory factors. The impossibility, he maintains, "appears to be rather a practical and perhaps temporary one, reflecting present limitations of knowledge of technology" (*59*, p. 371). The explanation *could* have been put in the deductive-nomological pattern *if* the information included in the explanation had been known and taken account of before the occurrence of the event. Possession of this information would have made possible the prediction of the event. Explanatory accounts that fall short of this standard he accords the status of partial explanation.

In other words, the failure of evolutionary explanations to be convertible into predictive arguments may be taken as a sign of their weakness or incompleteness. This weakness may reside in the conjectural nature of evolutionary speculations (especially when the only evidence for the conjecture is the fact to be explained), as in attempts to explain the occurrence of a rudimentary structure by postulating an ancestral function; or it may reside in the nonspecificity of evolutionary explanation, such that it may permit deduction merely that one of an indefinitely large class of adaptations to a given environmental situation will occur. Necessarily, according to this view, a full or complete explanation is one that is capable of yielding predictions.

On the other hand, it may not be entirely just to demean nonpredictive explanations by designating them as partial or imcomplete. As Scriven points out (*121*, pp. 479-80), the kind of predictability-in-principle that the Hempelian account requires is somewhat unhelpful, since its realization may depend upon our having sufficient foreknowledge to enable prediction of virtually everything. As long as we are unable to predict the earthquakes, ice ages, and volcanic eruptions in terms of which their consequences are explained, we shall continue to find ourselves explaining what we could not have predicted. Hempel is quite correct in maintaining that Scriven's argument does not show the logical impossibility of knowing all of the critical explanatory factors before or independently of the event to be explained. What is at issue, however, is rather whether these pragmatic distinctions ought to be reflected in our analyses of the notions of explanation and predictability. Given the fact that our ability to explain far outstrips our ability to predict, it does not seem unreasonable to expect this to be reflected in our decisions as to what sorts of accounts we choose to recognize as explanations. Those who deny explanatory status to nonpredictive accounts are left in the position of legislating, independently of actual linguistic practice, what is to count as an explanation.

Judged by the criterion of capacity to confer intelligibility, evolutionary accounts, if true, are indeed explanations, regardless of how "incomplete" they may be. The main thrust of the theory consists in its claim that it is the natural selection of variants and not special acts of creation or acquisition of inheritable characteristics by individual organisms that determines which types of organisms continue to survive, and that it is relative unfitness for the environment and not simply luck or accident that determines the differential birth and death rates among organisms. Evolutionary explanations purport to show that the nature of the earth's organisms, past and present, is in accordance with these principles. Every living organism represents the intersection of biological and environmental factors, the coming together of two causal pathways, one describable by the biology of inheritance and variation, the other by the physics and chemistry of its terrestrial and intragalactic environment. An evolutionary explanation typically indicates what factors have been causally relevant in determining the viability of a particular biological species. Hempel derogates such an account as telling the *story* of evolution, as distinct from providing explanatory insight into its processes. One can reply only that telling stories sometimes amounts to explaining.

I have argued that the fact that the theory of evolution often explains what it could not have predicted does not detract from the explanatory value of the theory. On the other hand, it also deserves to be noted that the theory is not entirely devoid of predictive power. Thus we are able to predict that the use of insecticides will result in the appearance of insect variants that are specifically resistant to these chemicals, even though we may not be able to predict which of an indefinitely large number of possible mechanisms will be developed to effect this resistance. Furthermore, there is the logical point to be made that the possibility of confirmation of a theory is in a sense tantamount to the possibility of prediction. Newly discovered biological facts are commonly explained according to evolutionary principles, but other such facts may serve to confirm the theory. It is not possible for the same fact to be used at the same time both as a confirming instance and as something to be explained. The difference is a pragmatic one: we seek to explain what we might find initially puzzling or unexpected, such as a peculiar structure or piece of behavior or a case of extinction; a confirmation is an act of discovering what could be expected to be the case if the theory were true, as in the discovery of certain fossils and "missing links." Every confirmation amounts to the validation of a possible prediction, since it involves the verification of a consequence of an antecedently entertained hypothesis. It is possible, of course, to have a theory which is never deliberately tested; such a theory would be confirmed only by its successes in providing explanations for additional facts as they are presented. Evolution is not such a theory, however, at least as it has been used to direct paleontological research. And if the investigation of fossils and the geological record involves the possibility of falsifying evolutionary hypotheses, then it must be capable of yielding confirmation as well. The theory of evolution can generate predictions if it can be confirmed, and it can be confirmed if it can be falsified.

At this point someone may ask what would falsify the theory of evolution. Indeed, the theory has been criticized as being so general as to be capable of explaining virtually anything. Since we can conceive of any organism as having descended from any other organism without rejecting the general theory, the theory has sometimes been characterized as nonfalsifiable.* And when

*See, e.g., *91*, pp. 64, 97; also Manser *84*. Nonfalsifiability would also follow, of course, if the theory is tautologous in nature; see note to p. 61.

evolutionary hypotheses involve positing selective advantages associated with every structure observed, it seems that the theory is compatible with any evidence whatever. Even cases in which acquired characteristics become part of a species' genetic equipment (for example, the "genetic assimilation" of enlarged anal papillae acquired by fruit flies grown on a salt medium (*154,* pp. 91-96) are apparently capable of being accommodated by the neo-Darwinian theory of natural selection.

Still, it is possible to conceive of occurrences which would be capable of significantly threatening the theory of evolution, if not actually falsifying it. Hayek (*57*) suggests the cases of horses suddenly giving birth to young with wings, and dogs giving birth to three-legged puppies as a result of cutting off the hind paw of successive generations of dogs. Such eventualities would at the very least force significant modifications upon the theory of random variation and natural selection. Another possible falsification of the theory as it is presently conceived might be the discovery of life on other planets such that the range of organisms corresponded to that on earth either at the present time or at some earlier stage of terrestrial natural history. It would appear fundamental to the neo-Darwinian theory of evolution not only that mutation occurs randomly, but also that the successful mutants that do appear represent only a small fraction of the number of biologically possible successful ones. The theory effectively precludes the possibility of reproductively isolated species evolving along identical pathways.

There is a more basic consideration, however, that concerns the falsifiability of the theory of evolution—or of any theory, for that matter. This is the point, originally made by Duhem, that, since a theory consists not of a single proposition but of an extensive network of

propositions and assumptions, the fundamental principles of a theory can always be maintained in the face of
any refractory evidence, providing we are willing to
make sufficient alterations elsewhere in the system.*
The history of science is rich with examples of efforts to
accommodate refractory experiences without abandon-
ing the fundamental principles of a theory, ranging from
Copernicus's assigning greater dimensions to the universe as a means of explaining his failure to observe
stellar parallax, to the phlogiston theorists' suggestion of
the negative weight of phlogiston to account for the
weight-gain of metals upon calcination, to the postulation of the neutrino (an undetectable particle having no
charge and no rest mass) to deflect an apparent challenge to the law of conservation of energy. Hypotheses
are testable only in groups, any one of which can be
considered falsified by a single piece of contrary evidence. The truth of any single hypothesis, whether it be
the central principle of a theory or a peripheral or auxiliary hypothesis, can always be maintained, either by
making appropriate additions to the propositions of the
theory or by rejecting some other hypothesis within the
group. It is for this reason that in the history of science,
as Kuhn has pointed out (*75*, pp. 149-50), theories are
never refuted; rather, their defenders die off.

There is a sense, therefore, in which nothing would
falsify the theory of evolution. No conceivable paleontological or geological finding is logically incapable of
being accommodated within the basic framework of the
neo-Darwinian theory, provided we are willing to make
sufficient adjustments in our theoretical net, either by
adding, deleting, or amending various auxiliary hypothe-

*34, pp. 184-90. See also Quine *110*, pp. 42-46, 77-79.

ses. Even the example of a horse suddenly giving birth
to young with wings, or a dog giving birth to three-
legged puppies as a result of cutting off the hind paw of
successive generations, might submit to explanation in
accordance with at least the basic outlines of the theory
of evolution. Indeed, evolutionists have posited large
quantum mutations ("macromutations") as a means of
explaining apparent "saltations" between disparate taxonomic groups,* and processes have been described (e.g.
genetic assimilation, referred to above) whereby artificially induced structural changes become hereditary
through natural selection. Even the discovery of apparently isolated yet qualitatively identical evolutionary
histories might be reconciled with the neo-Darwinian
theory, either by postulating heretofore unrecognized
prior and repeated mixing of the populations in question or by attempting to show that the constraints of
the particular environments have been such as uniquely
to determine which species survive. To be sure, some of
these accounts may be quite speculative and seem rather
farfetched, but the point remains that these accommo-
dations can always in principle be made and, in fact,
often are made by some, even when a theory has been
effectively supplanted. In this sense, theories may be
abandoned but never refuted.

To the extent that the theory of evolution, or any
other theory, is treated as unfalsifiable, it functions not
so much as a set of substantive or descriptive principles
as a set of regulative principles, as heuristic notions serving to guide the course of investigation. The neo-Darwinian theory of evolution tells us in what terms to
analyze the vast multiplicity of contingent biological

*See discussions by Mayr (*86*, pp. 435-38) and Grant (*49*, pp. 495-99).

phenomena we come across, whether in anatomy, pale-
ontology, physiology, or embryology. It guides us in
determining what to look for, such as evidence of an
organ's selective advantage or of a species' ancestral
types. If this regulative or heuristic character seems to
be more striking in the case of the theory of evolution
than in the case of other theories in the natural sciences,
I believe it is because of the fundamental looseness of
that theory: its capacity for encompassing a multiplicity
of phenomena and the relative ease with which it can
accommodate virtually any newly discovered biological
fact. This looseness also serves to explain why the task
of unseating the theory of evolution would be so enormous.

What alternatives might there be to a neo-Darwinian
theory of evolution by random variation and natural
selection? On the one hand, there have been various
positions which simply deny that species have evolved
from other species at all. The doctrine of the fixity of
species may be associated with a conception of the universe as static and eternal, or as involving numerous
separate acts of creation. On the other side are evolutionary theories which offer alternative mechanisms to
those described by Darwin and his successors. In addition to Lamarckism (the development of inheritable
characters in individual organisms through use and disuse), there are the theories which assert that the course
of evolution has been predetermined and fixed from the
start, and that evolutionary change is simply the unfolding of what was there from the beginning, or that evolution is directed by some goal or end toward which it is
moving. Such theories, whether framed in terms of the
imposition of an external plan or purpose or couched in
terms of an élan vital or an "inner dynamic of fulfill-

ment," specifically deny that the path evolution has taken would have been possible without such direction or orientation.*

This is not the place to undertake an examination and evaluation of these alternatives to the neo-Darwinian interpretation, but it is nevertheless interesting to note some of the implications of adopting such positions. Belief in a doctrine of multiple special creations, for example, should prepare one to expect literally anything, and can be expected to inhibit any search for connections between distinct species. Belief in a Lamarckian theory can have significant implications with regard to breeding experiments.† And if we believe that all evolutionary change is toward some end, we should be inclined to seek to discover final causes of individual biological phenomena and to expect subsequent events to provide a vindication of predictions based on the assignment of purposes. If it is a fundamental characteristic of a given species that it be in some sense striving toward a higher phylogenetic level, then a strictly mechanistic or behavioristic description of its present manifest features may miss the point entirely, in the same way that an account of individual human behavior in terms merely of physical movements will be irrelevant to the extent that it fails to indicate the purposiveness of such behavior. At the root of the problem of choosing between a theory of evolution and any of its competitors is the answer to the fundamental question of how we are to regard biological organisms.

The neo-Darwinian theory of evolution is an embodi-

*The literature in this area is prodigious. For references, see Hull 65.

†One of the most notorious examples of this sort in recent history was provided by the attempt to implement Lysenkoist genetics in the Soviet Union.

ment of the view that organisms and species resemble inanimate objects in that they come into being from other worldly objects, and that the laws that govern their generation, alteration, and disintegration are of a piece with those that concern other natural phenomena. The theory is the touchstone of modern biology because it provides a grounding for all of the biological sciences that is no less naturalistic and mechanistic than are the special fields themselves. And though the theory of evolution does, as I have argued, have certain testable implications, its major role is that of providing a framework for thinking about living matter, a framework that is not ordinarily challenged so long as it continues to prove fruitful in facilitating a systematic understanding of the vast domain of biological phenomena.

Explanation, Purpose, and Function

At the beginning of this chapter an explanation was characterized as a statement or set of statements that confers intelligibility on an event or state of affairs. We have seen that the biological sciences, like the physical sciences, typically achieve this end through the specification of causally relevant antecedent or concomitant circumstances and appropriate causal laws; an event is explained by citing its efficient cause. But there is another type of answer to "Why" questions that biologists find useful, namely, answers in terms of the function or purpose served by the structure or event in question. Why does a bee carry out its strange "dance"? In order to communicate the location of certain pollen-bearing plants. Why are human beings equipped with organs located in the inner ear known as semicircular canals? To provide a mechanism for maintaining balance. Why do locusts shed their exoskeletons? In order to allow for growth and replacement by larger ones. It would appear to be the case that such answers do give some degree of intellectual satisfaction and hence constitute a sort of explanation.

The fact that statements of this sort are found in biological writings at all suggests that they must be "scientifically respectable." The question arises whether functional or purposive statements in biology are analyzable in ordinary causal terms, or whether they are sui generis, perhaps demanding reference to special sorts of entities such as entelechies and vital principles, or embodying explanation by means of final causes. What is the relationship, for example, between such statements and the covering-law model of scientific explanation? Is biology committed to the use of a type of explanation radically different from any found in the physical sciences? In what sense, if any, do functional accounts explain the occurrence of the structures and processes to which they are applied?

Explanations that purport to answer "Why" questions by specifying a goal or end of which the object or event to be explained serves as a means of attaining are called "teleological." A teleological account, usually identifiable by its employment of such locutions as "in order that" and "for the sake of," is an attempt to explain present and past events in terms of their consequences, or rather their assumed consequences. Among the principle problems of teleology that philosophers of science have had to deal with are those of determining whether all teleological sentences (i.e. all sentences that employ teleological terms such as 'role,' 'purpose,' or 'function') are translatable into nonteleological sentences, and of accounting for the explanatory force of teleological notions in describing natural phenomena. There is a fundamental distinction between types of teleological statements, however. One class of teleological account deals with purposive behavior, where the attribution of purpose or intent may or may not be appropriate; the other is concerned with exhibiting the functions or roles played by things and events that occur within organized systems. The distinction appears to be that between an individual organism acting for a purpose and something being the case for a purpose. It is with only the latter of these two classes of teleological statement that our present concern lies; the former will be considered in chapter 4.

A functional explanation or analysis is an answer to the question, "What is the purpose of X?" In biology, X ranges over parts of organisms, processes and events occurring within organisms, and substances and suborganisms existing inside organisms. In general, a functional account is one that seeks to relate something to a system in which it participates.

The question, "What is the purpose of X?" may in many instances be ambiguous, however, for it may be construed as asking what purpose the designer of the system had in mind in creating the constituent in question, or it may be construed simply as asking what role that constituent plays in the system when it is functioning properly. Thus, in the case of an automobile engine or other human artifact, a particular structure may be an instrumental part in the operation of the machine, or it may merely be something the designer created to indicate that it is he who designed this particular machine. Typically, the purpose for which a given feature was created is to facilitate performance or maintenance of the system, although this need not always be the case.

Furthermore, even when the function of some element does happen to be the reason for its having been created, it will always be possible to distinguish a purpose from a function, if only because the former attaches to the producer, whereas the latter attaches to the product. What has a function may or may not have been created for that purpose, and what was created for a purpose may or may not actually serve the function for which it was intended.

Modern biology, of course, disdains explanation in terms of the wishes of a designer or creator. For a biologist, to ask the purpose of some feature of an organism is to seek to know its function, what it does for the organism or species in question. On the other hand, the biologist is also concerned with the question of how the organism came to possess that feature, or rather, how there came to be a species whose members bear that feature. It has been this consideration that has led to the attempt to define biological function in terms of evolutionary theory. Thus Canfield has proposed an analysis of function statements in biology according to which a specification of the function of a particular structure or process is equivalent to indicating how that feature is *useful* to its possessor, where "useful" is defined in terms of contributing either to the preservation of the life of the things that have it or to the maintenance of the species (*19*). What may look as though it were created by a designer with a purpose is thus explained as a consequence of nonteleological processes of natural selection of the results of random variation among the progeny of ancestral types. Those biological structures that have functions have them because the organisms in which they occur are the descendants of organisms whose ability to produce fertile offspring was dependent upon possessing those structures.

There is no question that evolutionary notions are relevant to understanding biological function. Evolutionary concepts are indeed valuable in bringing about understanding of why organisms possess certain features, and why certain functions are performed. Given not only that evolutionary explanations depend upon functional analyses to support the assumption, necessary for any neo-Darwinian account, that biological moieties do something useful for the organism or species, but also that evolutionary notions such as selective advantage and adaptive value provide significant clues for assigning functions to particular features, it is clear that there exists an integral and symbiotic relationship between functional analysis and the theory of evolution. On the other hand, as Frankfurt and Poole have pointed out (42), there is no need to suppose that evolutionary notions have to be constituents of the meanings of functional analyses. An analysis of function, or of functional explanation, is not the same as an analysis of why in general things have functions at all, or of how specific functional items came into existence. When I ask to be told the function of something, I may not care in the least how that thing came to exist. What I want to know is the use it serves; I expect a teleological reply. The attempt to replace a teleological statement with a nonteleological one fails because it amounts to proposing an answer to a substitute question. A request to know the final cause of something cannot be answered by supplying its efficient cause.

A number of philosophers, wishing to treat function statements nonteleologically, have tried to show that functional or teleological explanations fit the deductive-nomological model.* To give a functional account of a structure or process, it is argued, is to show how its presence is deducible from a combination of lawlike generalizations and statements of contingent fact concerning the effect for which it is a necessary condition. Thus the beating of the heart is functionally explained by deducing it from the statement that it is necessary for circulating the blood, which is necessary for survival. Similarly, the presence of chlorophyll is explained by saying that it is a necessary condition for the performance of photosynthesis, which is in turn necessary for the plant's self-maintenance. Functional explanations, by this account, are entirely on a par logically with ordinary causal explanations.

This position is open to two serious objections. The first is its inability to exclude statements of necessary conditions which have nothing to do with function. For example, the beating of the heart is a necessary condition for producing heart sounds, yet that is not the function of the beating of the heart, even though producing heart sounds is considered a necessary condition for being alive. A structure or process may have an effect which is of no importance to the functioning of the organism whose survival may still be a sufficient condition for its presence. And it cannot be satisfactory to reply that a given organism could in principle survive without producing heart sounds, but not without circulating the blood, since this same argument can be made for any internal process whatever, including blood circulation: oxygen might be transported by some other means. The notions of necessary conditions and deduc-

*See, e.g., Nagel *99*, pp. 401-28; Pap *104*, pp. 359-64. For a critical discussion of some of these attempts, see Hempel, "The Logic of Functional Analysis," in *59*, pp. 297-330; also Lehman *78*.

tive consequences are too weak to carry the burden of explicating function.

The other objection, related to the first, is that unless one adds the premiss that *only* the presence of the trait whose occurrence is in question could effect satisfaction of a particular need of the organism, no deductive-nomological argument is strong enough to explain why that trait, rather than one of its alternatives, should be present. As Hempel has pointed out (*59,* pp. 310, 313), such an argument allows us to infer merely that one or another of the items in the class of need-satisfying items exists. Furthermore, even if the deductive argument were valid, it would explain only the *occurrence* of the part or process in question, not its role or function. If functional accounts are explanatory, what they explain is what an item does, not simply that it exists.

The recognition that a functional explanation does not derive the existence of the item explained but rather shows the role that it plays in the organism as a whole suggests a radically different interpretation of function statements: if such explanations are to be understood in terms of the deductive-nomological model at all, per-haps the statement indicating the presence of the item whose function is in question should be regarded not as the *explanandum,* that which is to be explained, but rather as part of the *explanans,* the class of sentences adduced to account for the phenomenon to be ex-plained. In other words, to inquire as to the function of a part or process is to ask what process necessary to the survival of the organism is explained, in part, by citing its occurrence. Thus the item in question need not be presumed to be a *necessary* condition for the proper functioning of the organism or even of any process con-cerning it, but rather only as part of the sufficient conditions for proper functioning. What is explained in a functional account is not the item to which a function is ascribed but rather the functioning to which it makes an effective contribution.

According to this interpretation, to ask, "What does this thing do?" is like asking, "What can the occurrence of this thing be used to explain?" It is more like asking, "What does it cause?" than it is like "What causes it?" But it is not *exactly* like it. To state the function of something is not the same as to indicate its effect. For, to revert to an example used earlier, while it is true that the heartbeat has the effect of producing heart sounds, it is false that it has this function. We may explain, by indicating the cause of, heart sounds by pointing out the presence of a beating heart, but we are not in so doing giving any kind of functional account of a beating heart. What is lacking is the stipulation that the effect explained must satisfy some *need,* some condition the satisfaction of which is necessary for the proper functioning of the organism. A correct analysis or translation of the question, "What is the purpose of X?" then, is "Of what need-satisfying process would citing the presence of X provide partial explanation?"

Since the function of something is not the same as its effect, it must be acknowledged that a functional relation is not the same as a causal relation, although it necessarily includes one. The difference is that a functional relation is always either a three-termed relation, involving a structure or process, a process to which it contributes, and a system in which the other two elements occur, or a two-termed relation between the functional item and the system as a whole, whereas a causal relation is always a simple two-termed relation between two events. Furthermore, in a functional relation, the relation that is borne to the system must be one of contributing to its preservation or development. In biology the system is an individual organism or group of organisms whose survival is explicitly or implicitly referred to in a functional account. When it is noted that a function of the liver is the secretion of bile, the functional relation exhibited is not simply between the liver and the bile, but between the liver, the bile, and the whole animal, for which the secretion of bile by the liver satisfies a need. For an element to be functional, it is necessary but not sufficient that it be causally efficacious; to be attributed a function requires that its effect involves making a contribution to the maintenance of the organism or species as a whole.

Ascribing a function to something consists in determining what that thing does for the system with which it is associated. Ordinarily, that which is supposed to be benefited is readily identifiable; in biology it is typically the individual organism. In a number of instances, however, it is not the individual whose maintenance and preservation are promoted by a structure or process to which a function is ascribed, but rather a group of which the individual organism is only a single member. Furthermore, what may be useful to the group (e.g. a species) may have no utility at all for the individual or may even have negative utility. The process of childbearing in humans is an example of this; so is the loud call characteristic of birds that travel in large flocks, which serves to warn others of possible danger, but which could hardly be said to benefit the individual bird. Another area in which the performing of functions involves benefit not to individuals but to groups is the social sciences, wherein individual behavior is functionally analyzed in terms of benefit to the society in which the

individual lives. Functional analysis involves a preconception of the nature of the system that is being maintained, and it is this preconception that predetermines
what sort of functional account will be found acceptable.

The fact that functional analysis depends on prior
selection of the system that is supposed to be benefited
shows that the notion of function is essentially an extrinsic one. Before we can attribute a function to something, we must have determined what ultimate goal it
could be expected to serve. Usually, as in the case of
biology, we have a theory, such as the theory of evolution by natural selection, that provides us with this
conception and also guides us in the discovery of functional relationships. But the point is that the notion of
function must be brought *to* the situation that we wish
to describe in functional terms. To seek to know the
function of something presupposes knowing what having a function would consist of, and this depends on
having a preconception of what system that something
is supposed to serve, if it is indeed functional.

The conclusion we have drawn is basically a Kantian
one: the concept of function is not derived from the
phenomena we experience but is rather imposed upon
them as a necessary condition for the possibility of
experiencing them as comprehensible. Teleological judgment, according to Kant, views things of nature as serving one another as means to ends. Purposiveness, or objective finality, is attributed to that whose existence
"seems to presuppose the antecedent representation of
it" (*71*, p. 20). But, Kant argues, the existence of these
ends cannot be proved by experience. One can never
discover the adaptedness of one thing for another
through the study of nature alone. "We do not *observe*

the ends in nature as designed," Kant remarks. "We only *read* this conception *into* the facts as a guide to judgement in its reflection upon the products of nature" (*70*, p. 53).

Furthermore, a teleological principle, such as "Nature does nothing in vain," cannot be used to account for experience, according to Kant. Since such a principle is incapable of being established either a priori or by experience, it cannot serve as an *explanation* of any of the phenomena to which it is applied. The concept of purposiveness or of having a function is rather presupposed a priori as a necessary precondition for the possibility of ordering experience in a systematic manner. In Kant's terminology the concept of objective finality or purposiveness is not constitutive of nature but merely regulative; it "functions as a principle for conducting research" (*71*, p. 40). The concept of a thing as a "physical end" is used as "a regulative conception for guiding our investigation of objects of this kind by a remote analogy with our own causality according to ends generally" (*70*, p. 24) that is, our own purposive activity.

The interpretation of biological function that has been advanced in this section is essentially an extension or development of Kant's position. The ascription of a function to something is viewed as suggesting a purpose for which it might have been created. Seeking to know the function of something is tantamount to asking what need-satisfying process would be partially explained by citing the presence of the item in question; the pursuit of a functional account thus guides the search for a certain type of mechanistic explanation. The notion of function is treated as extrinsic to the system to which it is brought, and teleological principles are considered to be heuristic maxims whose justification resides in their fruitfulness in yielding systematic understanding of biological organisms and species. Functional accounts are useful because they embody a conceptual scheme that turns out to be applicable to biological systems, which, in Kant's phrase, form an "objective reality" to the conception of a teleological system.

Is there anything distinctively biological about explanations in biology? There is, of course, the trivial sense in which these explanations are distinctive in that they employ distinctively biological concepts, that is, concepts which happen to have application only to biological entities. But what is unique concerning biological explanation is not the logical type of the explanation, but rather the kind of thing that is to be explained. Biological explanations are distinctive because they concern self-regulating hierarchically organized open systems which are said to have evolved by random variation and natural selection, and whose constituent parts and processes are assumed to exhibit functional relations to the system as a whole. Biology explains how such systems work.

If organisms are arrangements of matter, and species are collections of organisms, past and present, then the things that organisms do are explained as effects of such accumulations of matter. Although a biological explanation may sometimes stop at a point short of microphysics or even biochemistry (for example, at the point of tracing neural pathways or of identifying the selective advantage of a given biological structure), there would seem to be no essential difference between intelligibility in biology and intelligibility in chemistry or physics. Only the paradigms are different, the phenomena that are treated as being themselves not in need of further explanation. No explanation needs to be viewed as ultimate in order to be acceptable, and since there is no reason to believe that any of biology's paradigms are incompatible with those of chemistry and physics, explanation in biology may be seen as distinctive only in its concern for certain levels and types of organization of matter.

3. Theories, Models, and the Concept of the Gene

If the purpose of science is to bring about understanding of the natural world, then the aim of biological science is to render intelligible the welter of phenomena experienced in the observation of living organisms. One of the things that biology, like every other science, does is to provide simple descriptions. It performs the role of exhibiting observed facts by expressing them in terms of concepts that have been developed as useful means of classifying a variety of objects that bear certain features by virtue of which they are designated as organisms. Once the descriptive categories or bases for classification are determined, the rendering of a scientific description becomes a matter of straightforward observation, classification, and application of predicates already contained within the scientist's lexicon. The process of description, in Peirce's terminology, consists in recognizing a certain thing as a token of a certain type, as a member of a certain class or a representative bearer of a certain property.

It is obvious, however, that not all scientific discourse consists of descriptive accounts of directly observable phenomena. Science (and prescientific speculation and mythology as well) has always taken as a point of departure the fact that some phenomena are found to be surprising, puzzling, or mystifying. Science exists because occurrences call for explanations and is moved by the tacit conviction that there is more to the natural world than immediately meets the eye. Scientific theories are one type of result of efforts to accommodate the panoply of natural phenomena, efforts to "make sense" of what is directly observed. Description alone, no matter how detailed, can never be expected to produce this intellectual satisfaction.

On the other hand, it is important to note the futility of attempting to maintain an ironclad distinction between description and explanation. There are at least two reasons why such a distinction is difficult, if not impossible, to draw. In the first place, in many cases a simple and direct description may actually be adequate to effect the degree of understanding desired and thus to provide a kind of explanation: a detailed account of how a human or an animal runs, or of how running differs from walking, might be an example of this type of description. Secondly, the fact that terms used in a direct observational description may be heavily "theory-laden" (meaning that they are defined only within a theory of the objects and properties they designate) shows that even the "mere" description of an object such as a cathode-ray tube or an internal-combustion engine may deserve to be counted as something of an explanation, assuming that the person for whom the description is intended understands the terms it contains. Accordingly, when we concern ourselves with the role of science in talking about things that are not directly observed, we need not commit ourselves, at least initially, to treating such accounts either as exclusively descriptive or as exclusively explanatory. Both functions of scientific discourse can be assumed to be operative in these areas.

In any event, every science, biology included, is engaged in attempting to indicate what else occurs in association with things that are directly observed to happen. Whether the phenomenon that evokes such an account is an earthquake, the apparent bending of a stick partially immersed in water, the curvature of the surface of a liquid contained in a thin vertical tube, the changes in ingested foodstuffs, or the arrangement of lines on an X-ray film, the move is to look beyond the observed features of the particular phenomenon in question and to try to relate it to other things and occurrences. Phenomena such as these are made intelligible by fitting them into systems in which other phenomena have already been ordered, either by subsuming them under general principles and hence assimilating them to other, less puzzling phenomena, or by explicating them as resultant effects of hidden mechanisms. In each case the phenomenon is placed under a theory whose principles are assumed to be capable of embracing more phenomena than merely the one under consideration.

Models, Analogies, and Theoretical Posits

An essential part of science, then, is the representation of what "lies behind" the phenomena of ordinary sense

experience. In some instances, what is said to "lie behind" does so only in an abstract or metaphorical sense:
the observed phenomena are treated as manifestations
of a plan or formal scheme. Examples of this type of
representation are the use of an optical diagram to account for an inverted image, the notion of lines of force
to account for the behavior of iron filings in the presence of a bar magnet, and the construction of a flow
diagram or neural net as a means of indicating what goes
on within an animal's nervous system. Sometimes the
representation is purely mathematical, in which case the
events observed are seen as the concrete realization of
relations expressed in mathematical equations. More
commonly, particularly outside of contemporary physics, that which is supposed to "lie behind" the phenomena is conceived of in mechanical (or paramechanical)
terms, or in terms of physicochemical principles that
have already found application in less unfamiliar domains. This type of account includes both cases in
which the states of affairs posited are subject to subsequent confirmation, such as the interior of a planet or
of an organism, and cases in which such direct confirmation is assumed to be impossible in principle, such as the
structure of the atom or the composition of a gas. The
creative aspect of science consists in bestowing order
and intelligibility on what is manifest through contributions such as these.

The means of representation that we have been talking
about are generally called models. Although the term
itself has been used so widely as to lack a single, precise
meaning, a model can be defined roughly as something
used to depict or otherwise exhibit coherently a class of
objects or occurrences for the purpose of achieving clari-
fication through simplification. It can be a physical

object—for example, a replica in miniature, as in the case of a model airplane or a scale model of the solar system or a geological formation. Another type of model is an abstract pattern or conceptual framework that provides a formal skeleton upon which to hang observed facts within a particular domain; economic models, such as the one which embodies the assumptions of perfect competition, represent examples of this sort, as do projected accounts of the ways mountain ranges are formed or religious institutions develop. Also included among models of this type are mathematical models of the sort used in modern physics. A third type of model involves the attribution of inner structure, composition, or mechanism to an object or system, and is usually formulated by analogy with things picturable in terms of familiar experience. The kinetic theory of gases, the Bohr model of the atom, and the Watson-Crick model of the DNA molecule are typical of this sort of model, and it is with this type that our present concern chiefly lies.

Models of the sort just mentioned will be called substantive theoretical models, as distinct from both formal and observable models. They are theoretical because they serve the same functions as theories, namely, explanation, prediction, systematization, and derivation of laws. They are substantive because they involve the imputation of structure. A theoretical model is the embodiment of a set of assumptions concerning what lies behind the directly observable. Models are not the same things as theories, however, despite the fact that the terms 'theory' and 'model' are often used interchangeably, as in the case of the Bohr model of the atom. A theory is a linguistic entity, a set of sentences, whereas a model is a postulated structure. The identification, therefore, at least so far as substantive theoretical models are concerned, is not between models and theories but between models and theoretical posits.*

Theoretical entities are objects posited as means of explaining what is observed. The type of inference according to which their existence is asserted was called by Peirce *abduction,* the devising of a hypothesis which, if true, would make the observed phenomena explicable as a matter of course.† Classically, there have been two principal positions with respect to the ontological status of these posits, that of the realist and that of the phenomenalist. For the former no ontological distinction is to be drawn between theoretical posits or constructs such as atoms and molecules and inferred entities such as pebbles inside rattling hubcaps and insects contained within Mexican jumping beans. For the operationalist or phenomenalist with respect to submicroscopic objects, on the other hand, much turns on whether a posited or inferred entity is held to be unobservable in fact or unobservable in principle: that which is believed to be unobservable in principle must be treated as a theoretical fiction. The difference in these positions has definite implications concerning the relation between theories and models. Certain conclusions bearing on this dispute as it concerns the status of theoretical constructs in biology will issue from what follows.

Before entering upon a consideration of the nature and uses of models in the biological sciences, however, it is important to attend to one further point of a general nature concerning models. A number of authors (e.g. Nagel *99,* chap. 6; Hesse *63*) look upon theoretical models essentially as analogies between the system the

*For further discussion of the relation between models and theories, see papers by Achinstein *1, 2;* Spector *136;* Swanson *142;* McMullin *87.*

†*105,* par. 145. See also Hanson *54,* chap. 4.

model is being devised to explain and a set of visualizable macroscopic objects, and models are sometimes spoken of simply as *analogs*. Bohr's atom is often represented as an analog of the solar system, chemical molecules as analogs of arrangements of balls and springs, light waves as analogs of ripples in water, and so on. To construct a model, on this view, is always to think of something as if it were something else.

Though it will be granted that models often are formulated and developed with the use of familiar analogies, however, it seems clearly to be a mistake always to treat models themselves as analogies. Suppose the atom is conceived as analogous to the solar system; this is not to say that it *is*, in any sense, a solar system, but merely that it is similar to the solar system in certain respects, such as consisting of objects of a determinate mass rotating in fixed orbits around a larger mass. What the atom is identified with is not a solar system but a system that is analogous to the solar system. To create a model of A on the analogy with B is to say, in effect, that A, like B, is an instance of C. The analogy is merely a convenient device for calling attention to certain features being attributed to the model that is being constructed.

At the very least, there is a fundamental ambiguity with respect to the relation between models and analogies. On the one hand, models, as conceptual devices, are constructed *by* analogies, not *as* analogies. On the other hand, sometimes the analog itself is physically constructed or assembled, in which case it is the analog that is generally called a model. Examples of this type include ball-and-stick models in organic chemistry, electronic models of portions of mammalian brains, and human-encounter models in social psychology. In these cases, what is identified as the model is obviously not the same thing as the object of the theory, but is supposed merely to embody certain features considered essential to the primary system. A model in this sense is, however, not a theoretical model but rather a model in the sense of a replica, like a model airplane. The point is not to deny that this is a perfectly respectable sense of the word 'model,' but only to show that it is quite different from the sense we have been exploiting, and that the identification of models and analogies depends on it.

Braithwaite, Toulmin, and others have advised us to beware of the danger of thinking of models as anything else than models, of identifying the objects of a model with the concepts of the theory. "Thinking of scientific theories by means of models," we are warned, "is always *as-if* thinking" (*16*, p. 93). "Models remain models, however far-reaching and fruitful their applications may become" (*149*, p. 167). If the kinds of models that these philosophers have in mind are not theoretical models but rather analogs, the point is an extremely trivial one. An analogy is obviously a relation between two *different* things; two things can be said to be analogous if they are similar in *some* respects, but not in *all* respects. If, on the other hand, it is theoretical models that are intended, the advice seems ill-taken. Why should it be *dangerous* to identify Bohr's atoms with systems of objects rotating in determined orbits around central masses, or gas molecules with elastic spheres? The reasons would appear to consist only in the possibility that these models are inaccurate or are oversimplifications. It is dangerous to make such characterizations, because one can turn out to be mistaken; but this danger exists with regard to any synthetic statement whatever. What these philosophers are concerned about is the danger of overdeveloping the theory by overextending an analogy. This danger, however, is none other than the danger of building a theory that outruns the evidence on which it is based, and has nothing to do with the methodological validity of treating theoretical models as theoretical posits.

Biology and Theoretical Concepts

The aim of the study which follows is to determine and assess the nature and function of theoretical models in biology. There is, as we have noted, considerably more to biology than anatomical description. Biology is indeed a theoretical subject, in the sense that it seeks to provide accounts of the observed in terms of the unobserved. The history of biology has been a history of theorizing, from the Hippocratic doctrine of the four humors to the speculative physiology of Galen and the modern physiology of Harvey and Bernard, to the development of contemporary biochemistry and biophysics. Whether their findings are presented in terms of as yet undiscovered structures, hypothetical constructs, or vital principles, biologists have consistently strived to present coherent accounts that go beyond mere description.

One way of illuminating biology's theoretical face is to examine in some detail, as a historical case-study, the development of one of its central theoretical concepts. Accordingly, we can expect an examination of that concept to yield philosophical fruits fully analogous to those that have come forth from the study of corresponding theoretical constructs in physics. The one theoretical concept that stands out from the ranks of those that have figured in modern biological writing is that of the gene. First, it is a fundamentally biological notion, having originally been posited to account for an exclusively biological process, namely, the inheritance of biological traits. Unlike the constructs of contemporary molecular biology, the gene-concept was not borrowed from chemistry or physics. Secondly, it forms the basis for what is systematically the most advanced and well-developed field of biological science. Classical genetics is the only portion of biology that has been completely axiomatized. Furthermore, though totally independent of theoretical physics and chemistry, the concept of the gene was developed as a strictly theoretical notion. Genetics has differed in this way from physiology, in which theorizing has always been assumed to await the discovery of an inferred structure or mechanism or isolation of a predicted substance. Finally, the gene as an inheritance factor has always been distinguished from posits such as vital principles (like Driesch's entelechies, which were supposed to be nonmaterial agencies which control embryonic development) by virtue of its explanatory power. To postulate a vital principle, whose own nature is inscrutable and unpredictable, is to concede that no explanation is possible; genes, on the other hand, were hypostatized as simple yet unseen factors whose behavior was supposed to be predictable and uncomplicated.

Genetics: The Mendelian Model

Before Mendel, the topic of biological inheritance was one about which descriptive knowledge was generally vague and theoretical speculation rather primitive. It was long known, of course, that individuals tend to resemble their parents. Aristotle, furthermore, had noted that a child may resemble a grandparent more closely than he does either parent. Inheritance was generally recognized to be biparental, even though as late as the eighteenth century it was typically assumed either that the matter of the offspring is contributed only by the mother, with the father contributing only the form or immaterial essence, or that heredity expresses merely the relative sexual forces of the parents. The prevailing view at the time that Mendel began his researches, in the middle of the nineteenth century, was that heredity was transmitted by an essence to which every organ in both parents contributed. This was the "blending" view of inheritance, and it is the one that was accepted by Darwin. Popular notions of "blood" heredity such as occasionally persist to the present derive from this conception.

On the other hand, it would be wrong to suppose that Mendel's findings sprang forth from a totally barren field. Like Newton, Darwin, and Lavoisier, Mendel had his predecessors, and his investigations were built upon their findings and speculations. In particular, there was the work on plant hybridization that had been carried on in the eighteenth and nineteenth centuries.* In 1760 Kölreuter had shown, for example, that intermediate hybrids do not breed true, but rather produce offspring of both grandparental types as well as of the intermediate parental type. Knight, Goss, and Seton, in the 1820s, all independently discovered dominance of seed color in garden pea hybrids, and also reappearance of

---

*For accounts of these and other aspects of the early history of the science of genetics discussed in this section, see books by Olby (*100*), Dunn (*35*), Singer (*132*), and Carlson (*20*); also articles by Glass (*43*), Grant (*48*), and Zirkle (*168*).

both parental colors in the next (F_2) generation. And Naudin, in the early 1860s, went so far as to postulate segregation of specific essences in the formation of germ cells in order to explain the observed reversion of the offspring of hybrids to parental types.

With all of the elements of simple single-factor inheritance known, there still was no theoretical comprehension of the process, for the simple reason that no one, prior to Mendel, had seen fit to put these elements together and to deduce testable implications. It was not sufficient to posit unseen factors so long as these could explain nothing beyond the phenomena they were devised to explain; Maupertuis (*85;* see also *44*) had done as much over a century earlier in conceiving the existence of a system of hereditary particles conveyed from generation to generation as a means of explaining certain hereditary conditions, such as albinism and polydactyly. Naudin saw that if hybrids form germ cells of the two parental types, there are three possible combinations, which would account for the observed three types of individuals found among the progeny of hybrid-hybrid crosses, but he failed to draw any implications concerning, for example, the relative numbers of these progeny. His hypothesis remained strictly hypothetical because he saw no way of evaluating it.

Mendel's contribution was essentially the result of his having undertaken to *count.* Only Mendel thought to draw inferences concerning the numbers of different forms that would result from the random fertilization of two kinds of egg cells by two kinds of pollen grain. In his 1865 paper he tells us that his aim is to discover a "generally applicable law governing the formation and development of hybrids" a task for which he deems it necessary "to determine the number of different forms under which the offspring of hybrids appear" and "to ascertain their statistical relations" (*89*, pp. 1-2). What Mendel did was to carry out an extensive series of breeding experiments in which he determined the nature and relative proportions of progeny resulting from crosses involving some twenty-two varieties of garden peas. By dealing with these results statistically, he was able to establish ratios such as could be predicted on the basis of a theory of segregation and independent assortment of character elements.

Specifically, Mendel found that whereas plants derived from a cross between two pure-breeding varieties were uniform with respect to the character being studied (e.g. round vs. wrinkled seeds) and resembled one of the parents (whose character was then called dominant), the next generation (the result of crossing hybrids with each other)—called the F_2 generation—always tended to a 3:1 ratio of parental types. Furthermore, whereas the individuals bearing the recessive trait were always found to breed true, the others were found to exhibit a proportion of 1:2 true-breeding dominants to hybrids that yielded progeny according to the 3:1 ratio. In addition, Mendel discovered that when two or three sets of inheritable characters were present, the various combinations (e.g. tall and round-seeded, dwarf and round-seeded, tall and wrinkle-seeded, dwarf and wrinkle-seeded) occurred in proportions predictable on the basis of the assumption that different pairs of traits behave in inheritance independently of one another. Finally, Mendel provided further confirmation of his theory by devising additional two-and three-factor crosses and by performing similar experiments using other species of plants (beans instead of peas).

It seems likely that Mendel has already arrived at his

theory by the time he performed the experiments described in his paper; otherwise it is difficult to explain why he set out to do those particular experiments. In any event, the results he obtained could have been deduced in advance from the particulate view of heredity whose principles they served to establish. The essence of Mendel's theory is the postulation of the segregation of character elements. If the character difference between the two forms of a given trait (for example, tall and short or round and wrinkled) is represented by the letters *A* and *a*, the resulting hybrid *Aa* will form two types of germ cell, *A* and *a*, so that the following types of fertilization can be expected to occur:

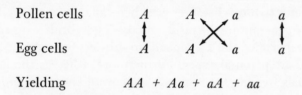

| Pollen cells | *A* | *A* | *a* | *a* |

Egg cells   *A*   *A*   *a*   *a*

Yielding   *AA* + *Aa* + *aA* + *aa*

It follows that the F_2 progeny should consist of pure dominants, hybrids, and pure recessives in a ratio of 1:2:1, which is what Mendel found. In the case of crosses involving multiple sets of characters, similar predictions could be made on the basis of assuming independent assortment. Thus Mendel asserted that in bifactorial inheritance four different types of germ cells are possible (*AB, Ab, aB, ab*) and that four types of individuals, in a 9:3:3:1 ratio, should result from crossing double hybrids with one another. His experimental findings were consistently in accord with these calculations.

Mendel's achievement consisted in providing a theoretical conception which not only rendered more intelli-

gible much of what was then already known about inheritance but also made possible the explanation of a
range of data that had not previously been recognized.
To the extent that he proposed the existence of an underlying order as a means of explaining observable patterns of inheritance, he created a model. Mendel's "elements" were hypothetical entities that determine the
presence or absence of specific characters in the resulting plants. "The differentiating characters of two
plants," he asserted, "can only depend upon differences
in the composition and grouping of the elements which
exist in the foundation cells" (*89*, p. 37). Mendel's inference was from the fact that similar hybrids yield dissimilar progeny to the conclusion that hybrids form different kinds of egg and pollen cells, from which he inferred the existence of formative elements that determine the characters in question.

Were Mendel's "elements" physiological units? The
answer that has usually been given to this question is
that they were not, that they were merely hypothetical
constructs devised solely for the purpose of rendering
systematic the distribution of observable characters in
inheritance. Mendel's interest lay in discovering laws of
inheritance, not in elucidating the fine structure of cells.
There is reason to believe that he considered the units
he postulated to be physiological in some sense or other
—in his classic paper on plant hybridization he indicates
his acceptance of the view that the development of an
organism from a fertilized ovum is determined by "the
material composition and arrangement of the elements
which meet in the cell in a vivifying union"—but he
apparently was not concerned with their nature as such
save insofar as was relevant for deducing the distribution
of character traits among successive generations of prog-

eny. Mendel's model was a very abstract one, and not one that depended upon or made use of any sort of material analogy.

It is interesting to contrast Mendel's conception with those of his more physiologically oriented contemporaries. Spencer (*137*, 1:180-83, 253-56), for example, though lacking a prediction-yielding theory of inheritance, had supposed the existence of physiological units, intermediate in size between cells and molecules, having the capacity to arrange themselves into special structures that serve both as transmitters of inheritance and directors of development. Darwin (*30*, chap. 27) postulated the existence of genetic particles called "gemmules" which were supposed to be generated by the body tissues and sent to the reproductive tissues by means of the circulating fluids. Similar ideas were put forward by Nägeli, who believed that heredity was transmitted by a substance called "idioplasm" which was carried by germ cells but also diffused throughout all the cells; by Weismann, who proposed a hierarchy of material bodies of ascending size ranging from "biophores" of the order of molecules to "idants" or chromosomes, which could be seen with a microscope; and by de Vries, who postulated a system of living, self-replicating, intracellular units called "pangenes" representing different hereditary predispositions. The multiplicity and diversity of these accounts is an indication of their basically speculative nature, with respect to which they may perhaps be compared with some of the seventeenth-century ideas of Buffon, Diderot, and Maupertuis. From a scientific point of view, what matters is not whether these accounts are correct or not, but whether and to what extent the evidence available would allow one to distin-

guish among them and to exclude other possible accounts.

If one's purpose in formulating a theory of inheritance
is to explain the regularities observed in plant hybridization experiments, it is not clear that speculation about
underlying physiological units is a significant part of
such a theory. Mendel, whose concern was in discovering general laws, was scarcely interested in describing
the material substrata of the processes he was trying to
systematize. Elucidation of hidden mechanisms can indeed be explanatory, but only if the posited structures
are assumed to have properties which serve to account
for the known properties of the system as a whole. The
trouble with proposals involving physiological units of
inheritance such as Weismann, Spencer, and Darwin postulated was that they could offer nothing but vague and
speculative accounts of the observed facts of development and regeneration. They were supposed to exist,
but they could be given no properties sufficiently definite to provide a mechanism of the processes in which
they were presumed to be involved. Mendel's conception, on the other hand, which involved units solely as
vehicles of transmission, cannot be faulted for its failure
to account for development. It was never expected to
play that role.

Mendel's model, if it is anything at all, is a *simple*
model. Inheritance is conceived as based on pairs of
particulate units, each of which determines a specific
trait. The makeup of each pair determines which of two
versions of its corresponding trait appears in its possessor, and each individual receives one or the other of
each pair from each parent. Finally, pairs of factors for
different traits assort independently in the formation of

egg and sperm cells, so that nonexclusive traits themselves will be found to assort independently. The reason Mendel was able to come upon such a simple model at all is that he was extraordinarily lucky. In particular, he was fortunate that he happened to choose as research material plants that exhibited the relatively simple patterns of inheritance he found. One of the significant disappointments of Mendel's life was his inability to extend his conclusions to the genus *Hieracium* (hawkweed), and it was the anomalous behavior of these plants in their inheritance that was at least partly responsible for the failure of Mendel's theory to gain acceptance during his lifetime. What is now called simple Mendelian inheritance is only one of a number of types of inheritance that have been elucidated since Mendel.

A simple model like Mendel's is valuable not only because it serves to render intelligible a welter of previously unordered data, but also because it is capable of being extended and because it provides a convenient object to be itself explained in terms of higher-level principles and chemical or physiological mechanisms. Before any such use could be made of the model, however, it was necessary for the theory to be accepted, or at least seriously considered. As it happened, Mendel's work was effectively neglected for thirty-five years, until the year 1900, when the principles were rediscovered almost simultaneously by three independent investigators. In the meantime, however, the way for acceptance of a particulate theory of inheritance was to some extent prepared by advances in the field of cytology, the microscopic study of the cell. Cells and cell nuclei had been described as early as 1831, but it was not until 1875 that Hertwig noted that the process of fertilization of an egg cell by a sperm cell involves the union of the two nuclei, an observation that led to the conclusion that the nucleus contains the physical basis for inheritance. As dissimilar as egg and sperm cells are to one another, they were found to contain similar parts which combine in fertilization, a fact that explained the similarity in the roles of the two parents in determining inheritance. In addition, it was discovered that cell division (i.e. the production of new cells, including germ cells) is accompanied by the longitudinal division of filamentous structures within the nucleus called chromosomes. By the end of the nineteenth century it was well established that egg and sperm cells contain just half the number of chromosomes that other cells do. A material basis had been provided for Mendel's implicit assumption that hereditary elements occur in pairs in body cells but singly in germ cells.

Prior to the rediscovery of Mendel's laws in 1900, however, no one appreciated the full significance of these cytological findings. Even Weismann, who not only identified the hitherto hypothetical genetic units with chromosomes but also recognized that reduction-division provided for the differential distribution of hereditary particles among the offspring, failed to see the relation between his conception and the Mendelian ratios, even after the rediscovery (*159, 43*). Those who were concerned with elucidating the microscopic mechanisms of intracellular processes apparently had little interest in creating a model that would account for the macroscopic effects that the plant hybridists were studying. The role played by cytology in the science of inheritance turned out to be merely the subservient one of providing confirmation of hypotheses which came forth out of other types of investigation.

The discovery of Mendel's classic paper and the rediscovery of Mendel's laws by three biologists in each case came as the result of experiments in plant hybridization such as Mendel himself had carried out. Only one of the three rediscoverers, Hugo de Vries, had previously been concerned with the material basis of inheritance, but it is safe to say that these speculations, prophetic as they turned out to be, failed to lead directly to the discovery of the Mendelian principles. In 1889 de Vries had published his *Intracellular Pangenesis* (*151*), in which he postulated submicroscopic, self-replicating units called "pangens" concentrated within the nucleus and representing separate hereditary characters. Lacking any conception of alternative forms or states of these hereditary units, de Vries was unable to derive laws of segregation from his theoretical model, however. Physiological units could very well be supposed to exist, but they were not known to *do* anything. It was only after de Vries was led into the field of experimental plant-breeding as a result of his speculation that alterations in these units (i.e. mutations) were responsible for the formation of new varieties and species that he began to comprehend the mechanism of inheritance. The discovery of the principles of segregation of hybrids depended on studying the behavior of the two forms of a hereditary unit on the level of observable features of undivided organisms. It was the breeding experiments (see *152*) that brought forth recognition of regularities for de Vries's physiological posits to explain.

The first decade of the twentieth century saw refinement of the Mendelian principles without significant alteration or delimitation of these patterns of inheritance. It was during this period that much of the terminology of what has since come to be called classical genetics was introduced, principally by Bateson and

Johannsen. What Mendel had referred to as *elements* and Bateson as *factors* were named *genes* by Johannsen. Genes were considered to exist in two forms, called *allelomorphs* (later shortened to *alleles*). *Zygotes*, or individuals produced in fertilization, were either *homozygotes*—formed by the union of two *gametes*, or germ cells, bearing the same allele—or *heterozygotes*—formed by the union of two gametes bearing different alleles. The *phenotype* of an individual, its appearance type, was distinguished from its *genotype*, which is constituted by its inherited predispositions or genes. The former is a purely descriptive notion, whereas the latter is theoretical, consisting of the set of invisible, discrete, separable units. This period also witnessed the elaboration of the gene concept as an experimental abstraction; the gene was left undefined and was to be used as a kind of accounting or calculating unit. As Johannsen summarized it, "The word 'gene' is completely free from any hypotheses; it expresses only the evident fact that, in any case, many characteristics of the organism are specified in the gametes by means of special conditions, foundations, and determiners which are present in unique, separate, and thereby independent ways" (*20*, pp. 20-22).

Even within the restrictions implied by this conception, however, there were controversies with respect to the character of these units. One of these concerned the notion of a "unit character." Does each hereditary character occur as a consequence of the inheritance of a single factor? Mendel, de Vries, and Bateson all more or less assumed that it does. Prior to Johannsen's introduction of the genotype-phenotype distinction, theoretical elements or factors and observable characters tended to be mentioned almost interchangeably. The vagueness persisted even after this distinction was made, however, since it was not made clear whether the observed character or phenotype was the result of the interaction of a single gene or of several genes with the environment. The matter was finally settled when it was shown that the full expression of certain characters may depend on not one but several pairs of independently segregating alleles: only 1/64 of the progeny resulting from a cross of hybrids produced from pure-breeding red and colorless-grained wheat plants were found to be colorless, a result to be expected only if three different gene pairs were involved (*35,* pp. 99-101).

Another issue that arose during this period concerned the question of whether there could be more than two alternative forms of the same gene. According to Bateson's presence-and-absence hypothesis, a recessive trait was to be represented as the loss of some element that is present in the dominant type. The discovery of multiple alleles, as in the case of full-colored, Himalayan albino (white with black points), and complete albino rabbits, indicated the possibility of alternative modifications of the same gene and effectively ruled out the more simplistic presence-and-absence hypothesis.

None of these ramifications represented a significant departure from Mendel's original conception, nor did they either strain or extend the model. By 1910 the gene concept had become well entrenched, and the new science of genetics (the name was coined by Bateson in 1906) was well under way. Although it was roundly acknowledged that chromosomes constitute the material basis of inheritance, the results of chromosome study were not yet an integral part of genetic theory, despite the fact that the distribution of pairs of chromosomes in the formation of germ cells had been shown to parallel the distribution of segregating pairs of alleles in Mendelian inheritance. The two fields of genetics and cytology had proceeded more or less independently, and neither had been used to extend the other. The situation was not unlike that which now exists with respect to psychology and neurophysiology: the latter is almost universally accepted to be the material basis for the former, and a number of correlations have been established, but there are very few if any cases of explanation of phenomena of one domain in terms of observations and concepts belonging to the other, except at a rather speculative level. Just as psychological states are believed to be associated with neurophysiological processes but are not defined in terms of them, so genes were associated with chromosomes but also not defined in terms thereof.

The Classical Gene

So long as genes were conceived simply as abstractions derived from breeding experiments, geneticists could not be expected to concern themselves with such things as chromosomes. In treating elements or genes as mere accounting or calculating units, Mendel and Johannsen left the concept free from any hypotheses concerning their mode of action or material composition. The chromosome theory of inheritance, developed mainly during the period from 1910 to 1939, was the outcome of a series of attempts to provide and confirm hypotheses of this sort. This was the direction in which efforts to extend the simple model of Mendel were taken.*

*Detailed accounts of the development and elaboration of the gene concept during this period (known as the "classical" period of genetics) have been given by Carlson (*20*), Dunn (*35*), and Whitehouse (*161*).

In examining how the gene model developed during this period, it is important to note that it was not cytological evidence that brought about acceptance of the chromosome theory, but rather genetic evidence against a background of cytological discoveries. Specifically, it was the discovery of inheritance ratios that departed from typical Mendelian ratios that turned geneticists toward controversies over intracellular mechanisms. The detailed story is quite an intricate one, especially if we consider some of the alternative proposals as well as the one that came to be accepted, but it is rather interesting as an illustration of the process whereby theoretical models are built up.

As early as 1902 Correns (*25*), one of the rediscoverers of Mendel, had found a case in which the Mendelian rule of independent assortment of separate genes appeared to be violated: two gene pairs had been investigated, one of which controlled white or blue color in corn seeds, the other of which determined self-fertility or self-sterility in these plants. Each character-pair, when studied independently, was seen to exhibit typical Mendelian ratios among the offspring of hybrids, namely, a 3:1 ratio of dominant to recessive phenotypes, white seeds and self-fertility being the dominant forms. When both sets of characters were tested simultaneously, however, the ratios expected on the basis of Mendelian principles were not found. Specifically, it was discovered that of the blue-seeded progeny *none* were self-fertile, whereas it would have been predicted that three-fourths of them would have been. Correns described the phenomenon he had observed as "coupling." The explanation he offered at the time was, basically, that the genetic makeup of the germ cells affects the ease with which they combine in fertilization; certain pollen cells tend not to form a union with certain egg cells. The kind of coupling Correns had in mind must have been at the phenotypic level (i.e. on the level of observed characters) rather than at the level of hypothetical particulate units, since he apparently never considered challenging the assumption that each hybrid plant produces all four possible types of gamete in equal numbers (gametes containing the dominant allele of both genes, gametes containing the recessive for both, and the two types containing one dominant and one recessive). Coupling was a purely descriptive notion for Correns.

Between 1905 and 1911 Bateson and others (see 7, pp. 148-53) reported additional cases involving coupling between distinct characters. It was discovered, for example, that among sweet pea plants of various colors the distribution of the color alleles was not random with respect to long and round pollen: instead of the expected 3:1 ratio of long to round regardless of color, a great excess of longs to rounds was found among the purples (about 12:1) and an excess of rounds among the reds (about 3:1), despite the fact that long is dominant to round. It is clear that Bateson could have accommodated these results to Correns's hypothesis of selective mating between gametes. Like Mendel, however, he was concerned to discover algebraic relations among his data and to provide an account that would go further than one which postulated differential mutual attraction between gametes of different genetic constitution. Bateson was able to show that the data he obtained could have been predicted if it were assumed that gametes containing factors for long pollen and blue color and gametes containing factors for round pollen and red color each occur with seven times the frequency of each of the other two possible combinations. In subsequent experiments involving different pairs of characters, he found evidence for a gametic ratio of the four respective combinations of 15:1:1:15. Bateson called this phenomenon *gametic coupling,* which he characterized as an association between the two dominant factors. Furthermore, he was able to interpret all of his data as indicative of geometric series of ratios according to the accompanying table, in which Aa and Bb represent the two pairs of alleles.

Gametes					Zygotes or individuals in F_2			
AB	Ab	aB	ab	$A\&B$	A only	B only	Neither A nor B	
1	1	1	1 = 4	9	3	3	1 =	16
3	1	1	3 = 8	41	7	7	9 =	64
7	1	1	7 = 16	177	15	15	49 =	256
15	1	1	15 = 32	737	31	31	225 =	1024
$n-1$	1	1	$n-1 = 2n$	$3n^2-(2n-1)$:	$2n-1$:	$2n-1$:	$n^2-(2n-1) =$	$4n^2$

Source: Bateson (7, p. 159); see also Carlson (20, p. 42).

In addition to coupling, Bateson discovered another type of departure from typical Mendelian ratios, which he called *spurious allelomorphism:* two dominant factors, instead of either segregating independently or associating with one another as in coupling, may in effect repel one another so that they are not both found in the same gamete (7, pp. 153-57). This phenomenon was characterized as a form of allelomorphism or allelism because it resembled the cases in which a pair of alternative factors always segregates from the hybrid; its spuriousness consisted in the fact that the two factors appeared to have no special physiological association. Later on the term spurious allelomorphism was replaced by the term 'repulsion.'

Bateson's initial explanation of spurious allelomorphism was that during the cell divisions of gametogenesis the dominant factors repelled each other. Coupling presented a greater problem, since it was necessary to account for the fact that unlike repulsion, which seemed to be always total, it was found to occur only partially. A few years later, however, the asymmetry of coupling and repulsion was disproved on the basis of the discovery of a single instance of an unexpected doubly recessive plant out of more than four hundred plants produced (8). According to the hypothesis of total repulsion, a hybrid should produce only two kinds of gametes, namely, *Ab* and *aB*—and the F_2 generation should consist of double dominants and the two types of single dominants in a ratio of 2:1:1. Using the same characters in sweet peas that they had used earlier in studies of coupling, Bateson and Punnett obtained a ratio of 226:95::97:1. Rather than regard the anomalous plant as an unaccountable mutation, they sought to explain it in terms of partial repulsion. On the basis of additional cases and deduction from other cases in which coupling had been found, they were able to draw up a table of repulsion ratios essentially analogous to the one on page 112. Repulsion and coupling could be viewed as converse phenomena.

By this time Bateson was able to propose a model by which he could explain these departures from Mendelian ratios. What was needed was a mechanism that would account for the production of gametes in the ratios he believed he had discovered. The geometric series of numbers suggested a geometrically ordered series of cell divisions in which alleles segregated according to certain patterns. Bateson called the process reduplication, and it was supposed to involve differential multiplication of certain combinations of alleles. The process whereby gametes could be formed in 3:1:1:3 and 7:1:1:7 is illustrated by the accompanying diagram.

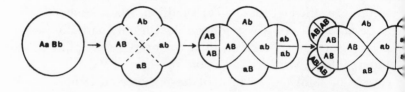

So long as the apparent gametic ratios were found to conform always to a geometric series, the reduplication hypothesis provided a plausible model. Unlike earlier conceptions of genetic factors, it was explicitly physiological, and although it claimed no specific support from cytology, it had definite implications in that domain. It was not cytology that provided the challenge to the hypothesis of reduplication, however, but rather genetics itself: violations of the geometrical series for partial coupling and repulsion. Ratios of phenotypes were discovered which failed to satisfy the ratios predicted by Bateson and Punnett. In a paper that is perhaps classic as an example of an effort to preserve a theory despite enormous strain, Trow (*150*) in 1913 reported and attempted to explain experiments which indicated gamete ratios of 2:1:1:2, 6:1:1:6, and finally $p:q:q:r$ (with no restrictions on the values of p and q) and even $p:q:r:s$, in terms of differential multiplication of cells.* In a passage which appears to be more of a boast than a concession, the author affirms that "the hypothesis of reduplication seems adequate to explain the occurrence

*The effective *coup de grace* to the reduplication theory was actually delivered the following year by Sturtevant (*140*), who showed that it required the postulation of astronomical numbers of successive cell divisions in order to account for the coupling and repulsion of five sets of linked genes in *Drosophila*, the fruit fly.

of any type of ratio" (p. 322). Clearly recognizing that the hypothesis rested on no evidence *other* than these ratios, he nevertheless concluded the paper by suggesting that the systems of segregation may prove to be of some value in the analysis of the structure of the protoplasm.

In the meantime, an alternative interpretation of these departures from Mendelian ratios was emerging, one which was based on contemporaneous knowledge and speculation concerning the chromosomes. Coupling and repulsion were to be explained in terms of linkage of genes contained within the same chromosome, and departures from complete linkage were to be explained in terms of material exchange or "crossing-over" of segments of paired chromosomes during gametogenesis. As obvious as this conception appears to anyone at all familiar with classical genetics, it is important to note the difficulty with which it was reached. It was not a model that immediately suggested itself upon collection of the relevant genetic and cytological data.

The ingredients of the synthesis that constituted Morgan's chromosome theory of heredity, which first appeared in print in 1911 (*93*), were cytological data on the segregation of chromosomes, Mendelian and non-Mendelian ratios of inheritance of simple characters, and the observation that homologous chromosomes twist around one another during the process of gamete formation. All but the last of these were recognized at least five years prior to the appearance of Morgan's theory, and Morgan himself published a lengthy review article in 1910 (*92*) in which he expressed serious doubts concerning the idea of locating genes on the chromosomes. Like Bateson, he had remained skeptical as to whether chromosomes were actually *directly* involved in inheritance, despite the repeated demonstrations of the parallelism between the distribution of pairs of chromosomes and the distribution of segregating pairs of alleles. Not even linkage or coupling, which was well known by that time and had been predicted by Sutton on the basis of a chromosome theory in 1903 (*141*) was sufficient to lead Morgan to accept a chromosome model of inheritance at that time.*

Morgan's early criticism of the chromosome theory is significant because it represents the rejection of a model by the man who subsequently was principally responsible for its nearly universal acceptance. Morgan had his reasons for rejecting the chromosome theory: there was no evidence that chromosomes separate after conjugation of homologous pairs along the line of fusion, nor was there evidence that inheritable characters "Mendelize" in groups commensurate with the number of chromosomes (an assumption necessitated by the fact that the number of such characters greatly exceeds the number of chromosome pairs), and the theory failed to account for the fact that the tissues and organs of an animal differ from each other despite the fact that all contain the same chromosome complex. There were plenty of grounds for being skeptical with respect to the chromosome theory, even though it was not in fact necessary for all of these objections to be fully overcome for Morgan to reverse his conviction only a year after he had posed them.

The most reasonable explanation for Morgan's apparently abrupt change of mind and sudden embrace of the

---

*It is interesting that Correns, who in 1902 not only discovered coupling but even advanced a theory which located the Mendelian factors in a serial order on the chromosomes (see also *26*), also failed to draw a connection between chromosomes and coupling.

chromosome theory seems to be that a relatively small increase in the amount of data to be accommodated by a theory convinced him of the need for a physical mechanism, one that did not exhibit the speculativeness and inelegance of Bateson's proposals in terms of coupling and repulsion. It could also be pointed out that Morgan's locating of the genetic materials in the chromosomes provided a means of simplifying the patterns of linking and segregation. Furthermore, his hypothesis had the advantage of *using* the data of cytology, however fragmentary, rather than disregarding it as irrelevant. The only new evidence that Morgan himself presented before his conversion to the chromosome theory was the discovery of a number of instances of sex-linked (or sex-limited, as he called them) characteristics in *Drosophila* (the fruit fly), which he reported simply as instances of coupling (*94*). Possibly the crucial stimulus came from Janssens's observations of twisting among paired chromosomes and the suggestion of an exchange or crossing-over of homologous segments upon the subsequent separation of the chromosomes. However, as Morgan himself acknowledged (*93*), the cytological evidence provided by Janssens was still not sufficient to prove that physical exchange did in fact occur, so it is clear that filling the gaps in the evidence cited by Morgan in his earlier review article was not essential to bringing about acceptance of the chromosome theory that he had previously been unwilling to accept. As obvious as the chromosome theory might appear, there had to be a large inferential leap for it to be adopted.

Morgan's conception of the gene was unequivocally material. What the chromosome theory amounted to was the explicit identification of the hypothetical factors posited by Mendel on the basis of breeding experi-

ments with physical constituents of observed microscopic bodies. Accordingly, subsequent elaboration and
confirmation of the theory involved both genetic and
cytological investigations. The genetic researches, led by
Morgan and his students, consisted largely in construct-
ing and testing chromosome "maps," whereby series of
genes could be arranged, on the basis of crossing-over
frequencies, in linear order on individual chromosomes.
That such mapping was possible represented a confirmation of the hypothesis that the frequency with which
linked or coupled genes separate is a measure of the
relative distance between them. Genetic investigations
also showed a correspondence between the number of
linkage groups and the number of pairs of homologous
chromosomes, and it was further shown in *Drosophila*
that the genetic map length for each of the four chromosomes corresponded roughly with their relative cytological lengths. This, plus the discovery that certain
chromosomal irregularities could be matched with corre-
spondingly anomalous genetic results, effectively
wrapped up the case for the chromosome theory, even
though it was not until many years later that it was
shown cytologically that genetic crossing-over is accompanied by interchange of actual segments of homologous chromosomes. The 1920s and 1930s saw considerable additional confirmation of the chromosome theory
through further cytological discoveries, but none of
these significantly altered the chromosome model.

A chromosome map represents the genes as lying in a
line—like beads on a string, as Morgan put it. The evidence for linearity was provided by the discovery that
for any three linked genes the map distance, or proportion of crossovers, between two of them is always approximately equal to the sum of the other two map

distances. But Morgan would go no further in attempting to characterize the genes themselves, beyond noting that they must have the power of self-division and of remaining unchanged through long periods of time, and that they seem to be of the order of magnitude of some of the larger chemical molecules (*95,* pp. 292, 320-21). Although the chromosomes were assumed to provide a physical mechanism for processes of inheritance, the genes themselves were not assumed to be observable entities. The principal difference between the "classical" gene of Morgan and his followers and that of the earlier Mendelians consists in the fact that the former conception carries with it a greater specificity with respect to the nature of the material basis of the units of biological inheritance. No one was seriously insisting that genes really *were* beads on strings, but only that they were analogous to them in certain respects, namely, their discreteness and colinearity.

What the chromosome theory of heredity achieved was the bringing together of two different lines of investigation, the genetic and the cytological. Mendel had discovered laws which enabled him to explain data concerning the inheritance of observable traits, and he and his followers posited subvisible elements or factors in order to explain these laws. Patterns of character distribution were explained in terms of the distribution of theoretical entities. Later, when departures from the typical Mendelian ratios of characters were found (i.e. coupling of unrelated characters), corresponding departures in the distribution of the theoretical units of inheritance were assumed. Concurrently, chromosomes had been observed to distribute themselves in ways that paralleled the distribution of the postulated factors. To suppose that these hereditary units are in fact parts of chromosomes was to offer an explanation not only of the way these units segregate but also of the way they sometimes fail to segregate. By explaining the postulated behavior of genes in terms of the observed behavior of chromosomes, the chromosome theory thus provided a further explanation of the inheritance patterns for which the gene model had originally been created. The chromosome theory explained why the gene model works.

The attempts to characterize the units of inheritance were not simply efforts to explain patterns of observed phenomena, however. As elsewhere in science, there was a definite concern with knowing structure for its own sake, whether or not such knowledge could at the time be used to explain anything. As has been argued in chapter 1, there is no science that is not descriptive as well as explanatory, so far as this distinction can be made at all. Every science includes a kind of natural history, whether it is concerned with describing geological formations, chemical molecules, or the atomic nucleus. Accordingly, the elaboration of a theoretical model can be a form of natural history, the providing of a description through indirect methods. Although a model may have initially been devised for the purpose of explaining, its subsequent development may be directed toward the specification of its structure.

Such was the case with respect to the experimental studies of gene mutation during the 1920s and 1930s. Muller's concern in launching this program was with understanding the properties and possibilities of the individual genes. Whereas investigation of linkage and of crossing-over had provided information concerning the boundaries of the units of inheritance, studies of apparent changes in these factors, it was believed, could yield theoretical knowledge concerning the specific nature and mode of action of the factors themselves. Research of this nature was not expected to explain anything at that time, however, but only to provide insights into the composition and organization of those parts of the cell that control inheritance, insights which might subsequently have explanatory value. The beads-on-a-string hypothesis was adequate to explain what was then known about the patterns of inherited characters; the mutation studies were concerned with elucidating the structure of the beads.

When de Vries introduced the term 'mutation' to the biological literature in 1901, it referred simply to the appearance of new forms of a species as a result of an abrupt change. When it was found that these character differences conformed to Mendelian patterns of inheritance, the term was transferred from the species to the hereditary units. Until 1913, when Sturtevant (*139*) demonstrated the existence of multiple alleles, showing that a gene could mutate in more than one way, it was generally assumed that a mutant gene was simply the absence of a factor that was present in the normal case; this was the "presence-and-absence" hypothesis. The discovery that a single character could have more than one alternative form determined by a single gene locus or set of alleles, plus the fact that mutations were found to occur "in reverse," effectively disposed of the view that mutation consists in the mere loss of genetic material from normal or wild-type individuals. In addition, it was pointed out, in a classic article by H. J. Muller (*97*), not only that there can be multiple allele series, but that the different alleles may affect totally different characters, suggesting that mutations may involve changes of radically different types within a given gene.

These were not the only inferences that Muller drew from studies of the distribution and frequency of mutations. He was able to show, for example, on the basis of finding that mutations occur only one at a time, and that the identical counterpart of a mutating gene on the homologous chromosome within the same cell remains unaffected, that mutations are spatially rather than chemically specific and are apparently due to "accidents" on a molecular scale. Muller also found that, although genes are in general quite stable—like radium, requiring more than a thousand years before half became changed—some are more mutable than others, and that mutation frequency itself is gene-controlled and can be altered by mutation at other loci. Finally, the discovery of differences in mutation rate at different temperatures, and of the effects of X-radiation on mutation frequencies, provided further bases for speculation on the structure of the gene.

Muller's extension of the gene concept was made possible by his taking seriously the assumption that genes exist as distinct substances, as particles too small to give a visible image. As a result of thinking of the gene as a material thing, he was able to ask and begin to answer a number of questions concerning what it does and what structure it must have in order to exert effects and undergo mutations in the way it does. Self-propagation, which Muller took to be the gene's most distinctive characteristic, was given an essentially chemical interpretation, for example, as a reaction in which some of the common surrounding material is converted into an end product identical with the original gene. Given the fact that this autocatalytic power is characteristically preserved through variation (i.e. mutation), he suggested that it derives from some universal structural feature such as a common chemical side chain. Similarly, the effects of temperature and radiation on mutation frequencies might be rendered more intelligible by viewing mutations as types of chemical reactions. Muller was even able to treat the discovery of "d'Hérelle bodies" (viruses)—ultramicroscopic self-propagating filterable particles capable of destroying bacteria—as evidence for the possibility of isolating individual genes. The corpuscular model had reached the stage of being capable of embracing data from a variety of areas of experimental inquiry.

Muller's assault on the problem of determining the structure of the individual gene marks a turning point in the history of genetics, in the sense that it represents the entry of a concern with physicochemical structure and mechanism into the biological domain. From a concern for the laws of the inheritance of traits within populations and for the intracellular processes involved in fertilization and cell division, geneticists moved toward a concern for the chemical makeup and mechanism of hereditary units. On the other hand, there was not at this time a corresponding change in the methods employed in investigating the processes of heredity. Breeding of individual organisms and classifying and counting the progeny continued to be the principal procedures of genetic research for several decades. Contemporary molecular genetics has come about essentially as a result of a series of attempts to elucidate the structure of Morgan's "beads," but the evidence that led to the alteration and ultimate renunciation of the classical conception was essentially genetic. The role of chemistry and physics was limited to that of a guide to reflection and speculation until at least the 1940s, as far as most geneticists were concerned.

The hypothesis of the chromosome as a string of beads enjoyed great success during the classical period of genetics. While incorporating all of the salient features of genes as then conceived, it was also able to accommodate multiple alleles and chromosomal irregularities (such as deletions and inversions of segments). Furthermore, it received support from the apparent heterogeneity of individual chromosomes observed during certain stages of cell division and from the discovery of visible bands on the "giant chromosomes" found in the salivary glands of *Drosophila* larvae. Genetic research had provided three primary ways of defining the gene: in terms of its effect on a particular character, in terms of recombination with other genes as a result of crossing-over, and in terms of its change to a different allelic form as a result of mutation. These were assumed to represent different aspects of the same structure, the one indicated by the string-of-beads analogy.

There was no evidence, however, that the units of function, of recombination, and of mutation were in fact coextensive, and by the mid-1920s evidence against the bead hypothesis had begun to accumulate. It was found, for example, that the appearance of an organism may be modified by the way supposedly discrete genes are distributed between homologous chromosomes, implying that the beads could not be conceived as isolated structures as had been supposed, but that neighboring beads (genes) could in effect influence one another, either directly or through interaction of the products of their activity.* Furthermore, the discovery of apparent

*This is known as the "position effect"; see, e.g., Dunn *35*, pp. 162-63, or Carlson *20*, chap. 13.

recombination between functionally similar mutants supposedly due to mutations in the same region of a chromosome linkage map indicated that the gene as a unit in physiological action is a larger chromosome segment than either the unit of mutation or of recombination. By the end of the 1950s separate and distinct definitions of the units of function, of mutation, and of recombination had to be given.

Viewed purely in genetic terms, the history of the gene model from the 1920s to the present has been a war between conservative defenders of genes as discrete, discontinuous units and radical attackers who would deny that there are any such things. Such a radical was R. B. Goldschmidt, who, recognizing that it was not a necessary consequence of the fact that genes can be defined at the functional level that they be differentiated from one another physically, suggested that it was the chromosome, which he thought of as a large chain molecule, that was the hereditary unit and not the gene at all (*46*, pp. 300-16). Mutations were regarded as changes in this molecule at specific points, changes that affected the rates of specific chemical reactions that determine the normal or wild-type form of the organism. As Whitehouse has put it (*161*, p. 173), Goldschmidt viewed the chromosome not as a string of beads but as a string without beads. His model was that of a continuum, like a violin string, in which different effects could be produced as a result of disturbances at various points along its length; in the chromosome these disturbances were assumed to take the form of alterations in the spatial configurations of the constituent atoms at specific points in the chain. Genes themselves were accorded no separate existence as individual units.

Goldschmidt's model was based on essentially the same evidence as was the classical gene theory, and was supported additionally by findings of the sort discussed above that led to alterations in the original beads-on-a-string model—evidence such as the discovery of inherited changes due to rearrangement of genetic material. The basic difference between the two models, as Goldschmidt himself pointed out (*45*), was a philosophical one: in place of an essentially atomistic conception which emphasized the functional discontinuity of the chromosomal material, Goldschmidt had offered a holistic account which stressed the role of the gene as part of a higher unit. Whereas the classical gene theory provided a simple and direct explanation of patterns of inheritance of specific traits, the continuum model never really attempted to explain the apparent discontinuities in the genetic material. On the other hand, the gene theory seemed to embody a kind of thinking that called for explanations in terms merely of additional genes of the same type, whereas Goldschmidt's model, by focusing on genic action and the catalytic activity of genetic material, suggested a biochemical approach to the problems of elucidating the mechanisms of inheritance.

Although Goldschmidt's theory was largely ignored by the supporters of the gene theory, it can be argued that the genetic model that emerged in the 1950s and 1960s bears as much similarity to Goldschmidt's conception as it does to the classical string-of-beads model. The contemporary molecular model, though it retains the identity of classical genes as self-replicating materially distinguishable functional units, no longer includes the assumptions that these units are indivisible, that they do not influence one another, or that the lines of demarcation between them, as defined by crossing-over, mutation, and function, coincide with one another. Furthermore, the current model incorporates Goldschmidt's assumption that there is no material distinction between the genes and the chromosomes, that the beads and the string connecting them are both made of the same material. The conservative defenders of the classical gene model of discrete, discontinuous units managed to win all of the individual battles against radical attackers who presented arguments against this conception, but in the winning of these battles the conservatives suffered casualties so extensive as to leave very little of the original conception intact. If genetics has never undergone a revolution, it certainly has submitted to radical reform.

Biochemical Genetics

Molecular or biochemical genetics can be said to have gotten its start when biologists began seriously to inquire as to the chemical nature of hereditary material, and to attempt to construct a chemical model of the gene. Although it has been known since the end of the last century that the chief constituent of the cell nucleus is nucleoprotein (a combination of nucleic acid and protein), nothing more specific was known concerning the chemical basis of heredity until 1944, when Avery, MacLeod, and McCarty (4) showed that deoxyribonucleic acid (DNA) is the substance that determines inheritance, at least in certain species. Since proteins were known to occur in a very large number of highly specific forms, whereas DNA had been found to yield upon hydrolysis only four distinct types of organic base, phosphoric acid, and a sugar, deoxyribose, it had long been assumed that genetic specificity resided in the protein part of the nucleoprotein, even though no attempt had been made to characterize the genetic material any further. It had been reported by Griffith in 1928 that material isolated from dead encapsulated strains of pneumonia bacteria was capable of permanently transforming avirulent (nonencapsulated) strains into virulent (encapsulated) ones, the possession of a polysaccharide capsule around the bacterial cell wall being a necessary condition for virulence; Avery and his associates found that DNA was the carrier of this specific hereditary character. Having identified a substance capable of transmitting a new hereditary property without sexual crossing, they also had provided a significant indication as to what goes on in ordinary processes of inheritance.*

Knowing the nature of the hereditary material is not the same as having a chemical model of the gene, however. In the first place, there was at this time no conception of the molecular structure of nucleic acid beyond that of the structure of its constituents. Secondly biochemists were still completely in the dark as to the mechanism of action and self-replication of the genetic material. It is therefore not surprising that the classically trained geneticists of the 1940s and early 1950s paid relatively little attention to the problem of elucidating the chemical basis of heredity.

For anyone interested in studying the self-replication of biological materials, phages—bacterial viruses—provided excellent experimental objects, since these particles (the d'Hérelle bodies referred to by Muller) had been shown to reproduce themselves by a factor of several hundred within a half hour following infection of a bacterial host cell. Research in this area was started in 1938 by Max Delbrück, a physicist who has indicated that he thought that biological processes might turn out

*Discussions of the origins and development of contemporary molecular genetics may be found in Ravin *112*, Carlson *20*, and Whitehouse *161*. See also articles by Stent (*138*) and Schaffner (*120*).

not to be wholly accountable in terms of conventional chemistry and physics, and hence was prepared to believe that the study of such processes might lead to the discovery of new physical principles, in the way that the investigation of subatomic phenomena had led to the discovery of new principles (*32*). Phage research, however, failed to yield any such principles. Neither did it provide a solution to the problem of self-replication in terms of known principles. What phage genetics did contribute was the elucidation of a system of autonomously self-reproducing entities, composed of nothing but protein and DNA, which exhibited both mutation and genetic recombination in a manner corresponding to chromosomal crossing-over in higher organisms. The culmination of this period of investigation came with the experimental demonstration in 1952 by Hershey and Chase (*61*) that it is only the DNA portion of the phage, and not the protein, that enters the bacterial host cell upon infection. Although this discovery, which showed DNA to be the genetic material of the phage, was essentially no more than an independent confirmation of the inference drawn by Avery, MacLeod, and McCarty eight years earlier, it played the important role of bringing into focus the double function of phage DNA as a genetic substance: DNA must not only be capable of reproducing itself but it must direct the synthesis of phage protein as well.

Meanwhile, other evidence of a more quantitative nature was being accumulated by biochemists concerning the distribution and composition of DNA. It was found, for example, in 1948, that whereas the amount of DNA contained in individual cells representing a variety of somatic tissues of the cow appeared to be constant, bovine sperm cells contained just half as much DNA, paralleling the distributions of chromosomes throughout the body. The amount of DNA per cell was also found to vary characteristically from species to species, and the intracellular doubling of the quantity of DNA was found to occur just prior to cell division during development (growth) and gametogenesis.

At approximately the same time as the discovery of the apparent species-specific character of the quantity of DNA per cell, it was found that the relative composition of an organism's DNA with respect to the four types of organic bases contained in DNA also exhibited such specificity. It was thus reported in 1949 that DNA isolated from different parts of a cow had the same proportions of the four nitrogen-containing bases adenine (A), guanine (G), cytosine (C), and thymine (T), and that these proportions were different from those characteristic of other species. From these and other data Chargaff (*22*) was able to derive a still more striking result, namely, that in every instance the number of adenine molecules (A) obtained as a result of the hydrolysis of the DNA was approximately equal to the number of thymine molecules (T), and likewise the number of guanine molecules (G) was close to that of cytosine molecules (C). These equivalences (which came to be called Chargaff's rules), apart from certain exceptions which could easily be accommodated, were found to be true of all species examined, although the proportion of the two base pairs (A-T, G-C) varied widely depending on the species.

It is important to note that Chargaff's inference depended not merely upon a quantitative approach to the problem of the nature of the hereditary substance, but also upon a type of thinking—"chemical thinking"—that involved deployment of a model that had not traditionally been used with respect to genetics. DNA was considered to represent one or more very large chemical molecules, and the various simpler substances isolated as a result of its destruction through chemical means were similarly viewed in molecular terms. Simple comparison of the relative percentages-by-weight of the types of organic bases obtained would not yield anything approaching Chargaff's rules. How does a chemist determine that two samples, weighing different amounts and representing chemically distinct substances, contain the same number of molecules? By comparing the respective rations of the weight of each substance to its molecular weight. (The molecular weight of any substance, so chemistry tells us, is the sum of the atomic weights of all of the individual atoms contained within a molecule. Chemical atomic weights, which are all computed relative to oxygen = 16, derive ultimately from measurements of relative combining weights of elements in simple compounds.) To say that A = T, or the number of adenine molecules is equal to the number of thymine molecules, is to say that the ratio of the weights of the two substances is equal to the ratio of their molecular weights.

Since neither Chargaff nor anyone else apparently had any notion as to the detailed structure of the DNA molecule as a whole, these biochemical findings were of no use to geneticists until Watson and Crick in 1953 proposed a model of the molecular structure of DNA (*158*), a model which would also yield a number of predictions concerning its biological behavior. The principal evidence upon which this model was based, apart from the sort of chemical knowledge that has already been cited, came from X-ray crystallographic studies carried out by Wilkins and his associates in the early 1950s. This technique, which produces diffraction patterns on film re-

flecting the symmetries and repeating structural features
of the crystalline materials to which it is applied, is
capable of yielding certain quantitative information concerning the spatial arrangement of atoms within a molecule. In the case of DNA the periodic spacings observed
in the X-ray diffraction pattern suggested a helix and
provided data which were interpreted as indicative of
the intramolecular spacing between successive turns in
the helix, the diameter of the helix, and the distance
between successive nucleotide units in the chain, each
such unit consisting of a base, a sugar, and a phosphate
group, all bonded together and to the next nucleotide in
the chain through ordinary chemical bonds. The key
feature of the Watson-Crick model was its postulation of
not one but two polynucleotide chains, coiled around
one another to form a double helix. The two chains
were supposed to be held together by hydrogen bonding—weak attractions known to exist between hydrogen
atoms and oxygen or nitrogen atoms in separate but
nearby molecules—between the bases, a single base on
one chain being hydrogen-bonded to a single base on the
other chain. Furthermore, this bonding was supposed to
be of a highly specific kind, such that only adenine and
thymine (A and T), and guanine and cytosine (G and C),
constituted the respective base pairs, thus accounting
for the equivalences found by Chargaff.

The way in which Watson and Crick proceeded, once
they became convinced from the X-ray photographs
that the DNA structure was essentially helical, was to
construct physical "Tinker-Toy" models out of components manufactured-to-order in a machine shop.* The
problem was to create a model in which the configura-

*For a stimulating account of the history of the working out of the DNA
structure, see Watson *156*.

tions of the atoms and the relative positions of the various atomic groupings were consistent with what was already known concerning the chemistry of DNA and with the X-ray data. As it turned out, the only structure that could be constructed which would fit all of the data available was one that involved specific pairing between the bases that Chargaff had found to occur in equivalent amounts. The structure, furthermore, imposed no restriction on either the base composition or the base sequence within a chain, so that, given the specificity of base-pairing between chains, the second chain would have a base sequence complementary to that of the first.

The significance of this molecular model, as Watson and Crick were quick to realize (*157, 158*), did not reside merely in its capacity to incorporate a welter of previously unorganized physical and chemical data. The payoff consisted in the possible biological implications of the postulated structure. One of these was the suggestion that the replication of genetic material occurs by the formation of new complementary chains by each of the two original DNA chains, following the breakage of the hydrogen bonds and unwinding and separation of the two chains, each chain acting as a template or mold for the formation of its complement. Another important aspect of the structure was its compatibility with any sequence of the four base pairs—A-T, T-A, G-C, and C-G—whatever, suggesting that it is the precise sequence of the bases in the nucleotide chain that determines the specific genetic effects. A third implication of the Watson-Crick model was that it provided a possible mechanism for genetic mutation: different configurational forms of the bases (called tautomers) known to occur with low but definite frequency could lead to "abnormal" base-pairing (for example, adenine with cytosine) and hence to "errors" in the formation of complementary chains, errors which might then be passed on to subsequent generations. In creating what a chemist or biochemist would call a chemical model with biological implications, Watson and Crick—a phage geneticist and an X-ray crystallographer, by training, respectively— could just as well be said to have devised a biological model that is also a chemical model, since it served to account for both classes of facts.

The Watson-Crick model was greeted with considerable initial skepticism on the part of many biologists, but it was not long before the proposed structure obtained sufficient support from subsequent experiments to gain widespread acceptance (see *112,* pp. 101-08; *161,* pp. 179-203). Examination of DNA under the electron microscope, for example, showed it to consist of elongated fibers of the specified diameter (even though it was not possible to determine whether these fibers were double-stranded and helical). Heating DNA at high temperatures was found to cause it to lose its rodlike shape rather abruptly, and this change was associated with approximately halving the molecular weight (suggesting a separation of the two strands). Further experiments were done (*90, 144*) which confirmed the hypothesis that replication would involve the synthesis of new DNA molecules consisting of one old and one new nucleotide chain. Another confirmation of certain essential features of the Watson-Crick model came from the test-tube synthesis of DNA by Kornberg and his associates in 1956, using the four nucleotides, inorganic salts, an enzyme, and a small amount of natural DNA as primer; the experiment showed that the nucleotide bases in DNA apparently *do* replicate by finding their components.*

The creation and subsequent predictive success of the Watson-Crick DNA model represented a significant triumph for the disciplines of organic chemistry and X-ray crystallography, in the sense that the model provided a demonstration of the applicability of their principles and techniques to a new domain. If there is a field for which this triumph represented a defeat, it is biology of the sort that resisted the notion that so fundamental a life process as the replication of hereditary substance could be explained in terms of ordinary chemistry and physics. Since the world seems to contain proportionally fewer and fewer biologists who retain that attitude, however, the DNA story can properly be viewed as a triumph for biology as well, at the expense only of ignorance.

From its inception, the gene-model has always constituted a point of tangency between two of biology's subsciences, genetics and embryology. The gene has *necessarily* occupied this role, for in serving to account for the inheritance of individual traits, it has been credited both with the transmission of these traits from preceding generations and with the determination of their development within individual organisms. Hereditary units have accordingly two distinct functions, that of self-reproduction and that of directing embryonic development. In contemporary terminology genes have been as-

*74. It is interesting to note that the Nobel prize was awarded to Kornberg for this work three years before Watson and Crick received theirs, despite the fact that the former could hardly have conceived of his experiments in the absence of the theoretical model contributed by the latter.

sumed to have both an autocatalytic function and a heterocatalytic one.

It is obvious that any adequate gene model must serve both developmental biology and genetics proper. As it happened, it was the attempt to solve the transmission problem that produced the contemporary model of the structure of the hereditary material, and it was the auto-catalytic function that this structure was first used to elucidate. Speculation and investigation of the hetero-catalytic function, the capacity of the gene to determine the control growth and metabolism within the living organism, however, were necessary before any progress could be made in terms of characterizing the genetic material at the level of discrete functional units that have traditionally been identified as genes. There had to be at least a rudimentary developmental biochemistry before there could be a full-blown molecular conception of the gene.

The question that needed to be answered was: What is the mechanism whereby genes control the development of hereditary traits? Long before there was a biochemi-cal model for gene replication, before there had been enunciated even a chromosome theory of heredity, the British physician Archibald Garrod in 1909 had suggest-ed that each of a number of inherited physiological de-fects ("inborn errors of metabolism") is due to a block or interruption at some point in a metabolic reaction sequence, and that the block was due to a deficiency of a specific enzyme. Garrod's work consisted in showing that the rare human metabolic disease alkaptonuria is inherited as if due to a single Mendelian recessive gene and is associated with the failure of the breakdown of certain amino acids as in normal metabolism, with the result that an abnormal substance is excreted in the urine. It was not until more than thirty years later that the influence of this work was felt in the development of genetics, when Beadle and Tatum in 1940 extended and generalized Garrod's discovery by demonstrating numerous cases in *Neurospora* (bread mold) of association between mutation of specific genes and blocking of specific biosynthetic steps, leading to the inference that every biochemical reaction is controlled by just one gene and through it by one enzyme. This assumption, known as the one-gene, one-enzyme hypothesis, led to the view that the gene's primary and possibly sole function was to impart a specific structure to an enzyme, thus endowing it with the peculiar ability to catalyze one step in a reaction sequence.

Since enzymes were known to be proteins, the problem of elucidating the heterocatalytic function of the gene (i.e. the mechanism of gene action) was considerably sharpened as a result of the introduction of the one-gene, one-enzyme hypothesis. The question became that of how genes direct the intracellular synthesis of proteins. More specifically, the molecular model posed the question of how chains of nucleotides determine the coming together of the various amino acids in specific ways to form chains that would constitute protein molecules.

During the time that it was generally assumed that it was the protein and not the nucleic acid portion of nucleoprotein that gave the gene its specificity, it was reasonable to believe that the genes were themselves enzymes and thus proteins. The transmission problem for a mechanistic science of inheritance would then have to involve a scheme for the self-replication of protein molecules, something about which no one had any insight. With the discovery that the hereditary substance is composed of DNA and the elaboration of the Watson-Crick structural model, with its built-in copying mechanism, came a relatively simple solution to the transmission problem. On the other hand, this discovery posed another problem which was previously not conceived to exist at all: the coding problem. There had to be some systematic way in which the structure of molecules that are not themselves proteins could determine the structure of molecules that are.

The coding problem arises because enzymes do not reproduce themselves and hence cannot be transmitted directly to an organism's progeny. Each individual must synthesize its own complement of enzymes anew from their amino acid constituents. If what an organism transmits to its offspring is simply DNA and exact or near-exact replicas thereof, then it is this DNA that determines which proteins are synthesized in the cells of the organisms that bear it. There must be, therefore, a definite correspondence between the type of DNA, determined by its sequence of bases, and the set of proteins characteristic of an organism. This correspondence is called coding, and the code is the set of rules according to which structural features of DNA are consistently correlated with structural features of protein. The coding problem is the problem of determining these rules.

The fundamental idea that turned the problem of elucidating the heterocatalytic function of genetic material into a coding problem was the proposition, variously called the sequence hypothesis, the colinearity hypothesis, and the theory of the genetic code, that the linear sequence of bases in nucleic acids is responsible for determining the linear sequence of amino acids in the polypeptide chains of protein molecules. (A protein molecule was known to consist of one or more chains of amino acids linked together in linear sequence. There are twenty different kinds of naturally occurring amino acids, and the three-dimensional structure of the protein molecule was assumed to be dependent on the specific sequence of amino acids within the polypeptide chains.) A specific base sequence could specify a particular type of protein by specifying its amino-acid sequence. To ask how it does so is, in part, to ask which base sequences determine which amino acids. Expressed in the material mode, this is to ask for the set of causal laws relating nucleotide sequences to amino-acid sequences; expressed in the formal or linguistic mode, it is to ask for the rules of translation between two languages, whose alphabets consist of the four bases and the twenty amino acids, respectively. The genetic code is the name that has been given to these correspondences.

This is not the place to relate the story of how the code has specifically been worked out, or to indicate what has been discovered concerning the mechanism of the intracellular synthesis of proteins within the cytoplasm in accordance with the nature of the DNA contained within the nucleus.* It is important to note, however, that these developments have all served as confirmation of what has come to be known as the "central dogma" of molecular genetics: that DNA carries out both the autocatalytic and heterocatalytic functions of the genetic material by serving as a template for the synthesis of replica polynucleotide chains; genetic "information" can be transmitted from nucleic acid to nucleic acid and from nucleic acid to protein, but never from protein to protein, or from protein to nucleic acid.

In addition, it should be pointed out that the investi-

*On this topic, see, e.g., Whitehouse *161,* chaps. 13-14; Woese *164;* Jukes *69,* chap. 2; Ravin *112,* chap. 5.

gation of the heterocatalytic function of genetic material through a study of the coding problem has led to further refinements of the gene-model itself.* The significance of the deciphering of the genetic code was that it provided a basis for a precise characterization of the gene in molecular terms. Mapping of genetic fine structure had produced a new definition of the gene as a functional unit, as distinguished from a unit of mutation or of recombination. The former unit had been given the name *cistron*, the latter two *muton* and *recon*, respectively. A cistron was shown to direct the synthesis of a single polypeptide chain, and hence the original one-gene, one-enzyme hypothesis had to be modified to a one-cistron, one-polypeptide hypothesis; the earlier idea of one gene specifying each enzyme is true only for enzymes containing only one kind of polypeptide chain. Since the solving of the genetic code has shown that each amino acid is specified uniquely by one or more of the sixty-four possible three-nucleotide sequences, a cistron that controls the synthesis of a peptide chain will consist of exactly three times as many nucleotide pairs as there are amino acids in the polypeptide it specifies. The muton and the recon, on the other hand, may be no larger than a single nucleotide pair, since a change in a single base, or the crossing-over of a single nucleotide pair between homologous chromosomes, could theoretically bring about a functional difference in the resulting individual.

Conclusions

The contemporary concept of the gene as a segment of a

*In particular, see Benzer's studies of phage mutants (*10*). For further discussion of this work and its implications, see Carlson *20*, chap. 22, and Ravin *112*, pp. 87-93, 108-10, 168.

DNA molecule obviously bears scant resemblance to the classical conception. On the other hand, the current molecular model is clearly a lineal descendant of the original Mendelian "elements." The history of the gene model has been coextensive with the history of the science of inheritance. At no time during this history was it correct to say that one model was *replaced* by another one; rather, the development appears to have been one involving successive correction and refining of earlier conceptions. There has been no revolution: geneticists have always succeeded in either repelling or accommodating critical attacks on the model. Even Goldschmidt, whose proposal that the gene concept be abandoned altogether was indeed revolutionary, and hence generally ignored, ultimately found certain quite radical aspects of his model (such as the idea that the chromosome is a large chain molecule and that mutations are changes in this molecule at specific points) incorporated into the DNA double helix model that came to be accepted by the "establishment." Like the sphericity of the earth and the heliocentricity of the solar system, the existence of genes as functional units describable in terms of ordinary chemistry is not a matter any contemporary biologist seriously doubts.

The concern of the science of heredity has been to order and describe the manifest facts of biological inheritance, to provide explanations of why these rather than other facts should obtain, and to indicate what lies behind them. Biology, like every other science, purports to tell us what there is. It becomes theoretical precisely to the extent to which the demands of order and intelligibility engender a need to impute nonobservable properties and substructures to the systems being described. The use of the gene-model has always been an instance of this. The reason why no adequate explanation is possible without resorting to such a model is not unlike the reason why there can be no explanation of an automobile's backfiring without referring to the engine and its constituent parts or of lightning without mentioning electrical charge: part of what we mean by "explain" in the case of a physical occurrence is that molar phenomena are subject to analysis in terms of microscopic or submicroscopic elements.

One of the most important results of our study of the gene is the finding that it is impossible to maintain, in biology, at least, any sort of radical distinction between inferred entities and logical constructs. The essential difference between those who, like Mendel and Johannsen, limited themselves to strictly operational conceptions of the gene, and those who, like Bateson, Morgan, and the molecular geneticists, were fully prepared to think of the gene as a physiological unit, appears to reside in their respective degrees of conservatism as to how far they were willing to go in attempting to characterize the hereditary material. There is no clear distinction, either logical or ontological, between a definition of something in terms of its particular nature and one in terms of what it does. As far as the biologists themselves were concerned, to have been skeptical of the model at any point was to have doubted its accuracy, not the existence of some sort of model altogether.

Investigation of the history of the gene-model has also revealed the dubiousness of distinguishing between what is said to be unobservable in fact and what is said to be unobservable in principle. The ambiguity of cytology's role in producing evidence through microscopic examination of the cell nucleus for the segregation of hereditary elements according to Mendelian principles is an illustration of the difficulty of drawing such a distinction, as is that of the use of microscopic techniques to demonstrate the inversion of chromosomal segments as confirmation of the chromosome theory of heredity. More recently, photographs taken with the electron microscope have provided observational evidence for the existence of large chemical molecules of the general shape suggested in the construction of biochemical models of the hereditary substance. For the geneticist, a theoretical construct seems always to have been an inferred entity; the role of direct or indirect observation is simply that of confirming a theoretical prediction, much as the first telescopic observation of the planet Neptune provided confirmation of the prediction of Adams and Leverrier based on Newton's theoretical mechanics. Given the existence of a spectrum of objects ranging from microscopes and microorganisms to macromolecules and atoms, one is compelled to concede the impossibility of making a metaphysical distinction between the observable and the unobservable, and with it the untenability of any position, such as instrumentalism or operationism, which treats theoretical objects as fictions and observable objects as real.

For many biologists the successes of molecular genetics in explaining how traits are inherited has represented proof of the concrete reality of a model that had previously been viewed as a strictly biological construct. According to this conception, chemistry is basic, and its models are taken to have the ontological status of microscopic and macroscopic entities. It also follows that biology, or at least genetics, must be regarded as incapable of yielding a *complete* explanation of the phenomena that come within its purview; if the ultimate principles appealed to are not those of chemistry and physics and hence essentially *non*biological, the explanation cannot be regarded as complete. Biology proper, with its own concepts, principles, and models, is thus conceived as a means of arranging and packaging objects within a certain domain, ready for either providing answers to biological questions or for further analysis in terms of nonbiological constituents.

As a theoretical model, on the other hand, there is logically no need that a biological posit be grounded in anything outside of biology in order to be explanatory, any more than there is a need for the constructs of psychoanalytic theory or behavioral psychology to be identified with neurophysiological or other physico-chemical constructs in order to have explanatory and predictive value. As it happens, in the case of contemporary genetics and molecular biology the identifications have to a considerable extent been made and the models can be said to have coalesced. Since the compatibility of a strictly biological model of the gene is almost never at issue, however—even those biologists who emphasize most vehemently the uniqueness of biological systems are typically prepared to acknowledge that this uniqueness is attributable to the fact that these systems simply happen to exhibit configurations and structural organization of a kind not found in nonbiological systems—the gene can still be construed in strictly biological terms and yet be intelligibly invoked as a means of explaining a number of biological phenomena. It would be a mistake, in any case, to deny that molecular models are any less "theoretical" than purely biological ones are; the difference between them lies in the generality of the theories by which they are deployed. If a chemical model is deemed to have a privileged status relative to a biological one, it is only because of the role it plays in providing a substratum for the biological model.

Chemistry can be seen as the hard money backing the paper currency of biological statements. A tentative or provisional model is a form of currency whose backing (ultimate worth) may still be in doubt. It may be intentionally artificial, as in the use of a Ptolemaic model in celestial navigation (corresponding to the use of scrip or chips), or may be acceptable as legal tender, as in the biological sciences. Observations are the goods whose production and delivery are obtainable with the use of currency, and like the negotiability of currency where gold is not directly employable, biological statements have a utility in transactions in which physicochemical principles are not appropriate. But the fact that it is often neither convenient nor profitable to exchange one's currency for that which has a wider range of applicability does not contravene its commensurability.

If any area within biological science has a claim to reducibility, it is classical genetics. There is no known biological fact concerning the transmission of hereditary traits for which there is not at least strong evidence of its explicability in terms of chemistry. When the explanation of the reproduction of genetic material has been given at the molecular level, genetics as a purely biological science can be said to have been completed, save for the subsequent working out of further details and refinements. The birth of molecular genetics (which would have been literally impossible were it not for the techniques and discoveries of the classical disciplines) has marked the closing of a chapter in the history of science. Scientific histories can be closed in a way that human history cannot, because the basic questions that are answered in the course of the unfolding of such a history are asked at the outset. Genetics is complete in the sense that it has brought us to the point where we can say, "Now we know." This is not to say we know *everything* about inheritance—the history of knowledge need not presuppose an ultimate end at all—but only to say that we know everything we wanted to know when we originally asked how organisms transmit their likeness to subsequent generations.

4. The Presuppositions of Biological Science

Science is concerned with the behavior of things. The reasonableness of this statement is the result of the ambiguity of the word 'behavior,' whose use has been broadened so that it has come to be applied not only to humans but to organisms in general, and finally to entities of any sort whatever. One can speak of the behavior of a child, of the behavior of a rat in a cage, and of the behavior of steel under stress. Biology exhibits a variety of ways in which objects are said to "behave": biologists may find themselves talking about the behavior of bees inside the hive, the behavior of the mammalian heart under digitalis, and the behavior of DNA in the presence of 5-fluorouracil. Biology concerns behavior on a number of levels.

The distinctiveness of biology resides not so much in the fact that it embraces a multiplicity of levels, however, as it does in the apparent uniqueness of one of its levels, namely, that of the whole organism. Explorations carried out on lower levels are considered part of biology only to the extent that they are expected to contribute to an understanding of the organisms in which they occur. Biological science, furthermore, will be found to involve characteristic paradigms and presuppositions which incorporate a priori notions concerning its standards of explanatory adequacy and the essential nature of its subject matter. Investigating what is peculiar to biology requires uncovering and scrutinizing its implicit assumptions and the attitudes they embody. The results of such investigation should serve to illuminate not only the demands biological materials make on the nature of biological science but also the restrictions that biological science places on those aspects and segments of the world that come within its purview.

Organisms as Machines

One of the most pervasive of the presuppositions of biological research has been that organisms are to be regarded as physical mechanisms. It has also been a topic of considerable philosophical controversy, controversy that has to some extent been dependent upon what the words 'physical' and 'mechanism' are taken to mean. An organism is obviously not a mechanism in the sense that a clock is; but then, neither is an electronic computer. Physics itself has become less "mechanistic," in the sense that it has moved away from its earlier Newtonian paradigms. Even with respect to inanimate matter, it is now recognized that higher orders may not be completely determined by the lower-order mechanical motions of particles. As the physicist David Bohm has remarked, it is a curious feature of modern biology that it is moving closer to mechanism as physics is moving farther away from it (*155*, 2:34).

The thesis that organisms are "nothing but" physical mechanisms has had a long history of opposition by distinguished biologists, even down to the present day. Some of the criticisms have been vitiated by subsequent empirical discoveries, but others can be seen to represent a real challenge to prevailing contemporary attitudes toward biological science. With respect to the former type of dissent there have been not only those of the vitalists of the nineteenth century and earlier, but also the attacks of antivitalists of our own century, such as J. S. Haldane, who wrote that "the phenomena of life are of such a nature that no physical or chemical explanation of them is remotely conceivable" (*52*, p. 64). In particular, this biologist (whose contributions to physiology have guaranteed him a prominent place in the history of biology of the early twentieth and late nineteenth centuries) believed that no theory of heredity along mechanistic lines was possible, and wrote as late as 1931 that mechanistic theories of reproduction "never make anything else than sheer nonsense" (*53*, p. 147). The fact that subsequent research has shown Haldane's pronouncements to have been somewhat ill-taken is an illustration of the point that scientific advances can sometimes supersede declarations that have the ring of metaphysical conclusions.

Other attacks on the mechanistic position (if that is what the thesis that organisms are physical mechanisms is to be called) have focused on the salient differences between organisms and machines. Woodger thus lists five fundamental respects in which organisms differ from machines: the fact that their parts have different properties when separated from the whole from what they have when they are not separated, that they are able to adapt to changes in the environment, that they are not known to be dependent for their existence on any human mind or any other mind, that they represent the outcome of evolutionary processes (in the biological sense), and that they are genetically related to one another (*167,* pp. 451-52). But even if one accepts all of these propositions, it is by no means clear that one should be led to the conclusion that an organism is not a physical mechanism, for it may be replied that only a rather narrow and rigid notion of mechanism would be refuted by these arguments. It has never been the point of a mechanistic position to attempt to assimilate organisms to existing hunks of hardware or even to ones projected on drawing boards. The question is not whether there can be, for example, a machine that grows in the way an organism does, but whether the organisms can be understood as a machine that grows, in a non-vitalistic sense of "grows."

Another attempt to discredit the "machine theory" of biological organization has been made by von Bertalanffy, who pointed out that "every machine is where and what it is for a definite purpose, and that it presupposes the engineer who has conceived and constructed it" (*12,* pp. 37-38). What we have in biology, he insists, is not a machine theory but a machine fiction—a fiction that allows us to say that organisms can be regarded *as if* they were machines. Since the fundamental problem of biology is, for von Bertalanffy, the organization and *self*-regulation of materials and processes, he does not think that this is even a useful fiction.

Von Bertalanffy's position is not ultimately incompatible with regarding organisms as physical mechanisms, however. The inference he draws after noting the inadequacy of treating the organisms as a closed system capable of maintaining complete chemical equilibrium is not that biological systems fall outside the domain of physics and chemistry, but rather that they must be viewed as *open* systems that are never in true equilibrium but maintain themselves in or near a steady state. A living organism, he affirms, is a hierarchical order of open systems which maintains itself in the exchange of components by virtue of its system conditions (*13*, p. 129). As such, it is clearly a kind of physical system, and the science that is needed to describe it is considered to be *a new field of physics*. Biological systems, far from being treated as essentially nonphysical systems, are regarded as examples of physical systems of a general type that demand only an expansion of standard kinetics and thermodynamics to take into account the fact that every living organism continually gives up matter to the outer world and takes in matter from it.

On the other hand, as eminent a physicist as Niels Bohr has suggested that, because the conditions holding for biological and physical researches are not directly comparable, it may be necessary to take the existence of life as an elementary fact that cannot be explained (*15*). Arguing by analogy with the insufficiency of mechanical analysis for an understanding of the stability of atoms, Bohr considered it reasonable to believe that an understanding of biological phenomena would ultimately depend on the recognition of some fundamental complementary relation holding for living aggregates of matter. These ideas were extended by Max Delbrück, who, as we have seen (pp. 128-29, above), thought that the investigation of the living cell, particularly with respect to genetic replication, might reveal certain fundamental paradoxes that would signify natural limits to the approach of molecular physics, analogous to the finding that, unless the notion of stable orbits and quantum jumps between them is brought in, no system consisting simply of electrons and a nucleus could have the properties that atoms were known to have. Because of this apparent complementarity between the living cell and atomic physics, Delbrück urged that the analysis of the behavior of living cells be carried out on the cell's own terms without fear of contradicting molecular physics.

As far as the molecular biology of inheritance has turned out, however, no paradoxes have emerged, and there is, if anything, less indication that biological research is likely to turn any up. Furthermore, as Bohr and Delbrück both were well aware, even if subsequent investigation does lead to the formulation of paradoxes and new complementarity relations, what these would yield would be essentially new laws of physics. Their existence would show not that biological systems elude or contradict the laws of physics, but only that the "old" physics is inadequate. It would show only the naïveté of assuming organisms to be physical mechanisms in an archaic sense of mechanism.

Biologists have sometimes urged an attitude of "open-mindedness" (see, e.g., *133*, p. 59) with respect to the problem of how the phenomena of organic self-regulation are to be explicated. But what does it mean to be open-minded in this instance? It means, presumably, to be willing to consider any of the following three possibilities: organic self-regulation is ultimately explicable using (a) known physical principles, (b) to-be-discovered physical principles, or (c) nonphysical principles. The first two of these possibilities are clearly compatible with the thesis that organisms are physical mechanisms, given a liberal interpretation of that thesis. If the third possibility obtains, however, either organisms are governed by nonphysical principles that exhibit no invariant relationships to particular configurations of matter, or else biological matter possesses characteristics that structurally identical or analogous nonbiological aggregates would not. The former is essentially spiritualism, the latter vitalism. If the falsity of both of these positions is a presupposition of modern biology, then the counsel of open-mindedness must be construed only as advising biologists to be prepared to be occasionally unsuccessful in their efforts to elucidate the phenomena of living systems in terms of known principles of chemistry and physics.

Organisms, Machines, and Emergence

If there is an alternative to the view that organisms are physical mechanisms, it is the position that living systems constitute domains in which the laws of chemistry and physics do not hold. The progress of biochemistry and biophysics notwithstanding, there is no way of proving that the chemistry and physics of biological systems is in all respects the same as it is elsewhere. On the other hand, it may be asked whether any contemporary biologist would be willing to accept the antimechanist position, stated thus baldly. As has already been acknowledged, organisms are clearly not machines of the rather simple sort that antimechanists typically imagine in attacking the "machine theory," but that fact alone does not detract from the role and the plausibility of the assumption that organisms are to be conceived as physical mechanisms.

In order for the assumption that a living organism is

essentially a naturally occurring machine (albeit one which cannot be understood without also considering its interaction with its surroundings) to be of any use to biologists in achieving an understanding of their subject, it must also be assumed, of course, that machines are the sorts of things we can understand. For most philosophers, to be able to treat anything as a machine is tantamount to regarding it as explicable in terms of chemistry and physics. We have seen that there is at least one philosopher who considers biological systems to be inexplicable in physicochemical terms just precisely because they *are* physical mechanisms: Michael Polanyi. According to Polanyi (*108*, pp. 41-42), the fact that it is not possible to derive the structure of any machine or its operational principles from the principles of physics and chemistry alone shows that it is wrong to suppose that a mechanical explanation of living functions amounts to their explanation in terms of physics and chemistry. As I have argued in chapter 1, however, what Polanyi's argument shows is merely that, owing to the large number of arrangements of materials all equally consistent with physicochemical theory, the initial configuration of elements is taken as given, as initial conditions to be specified as part of a physicochemical explanation. The nonderivability of the principles of higher levels of organization from lower-level principles signifies not that the higher levels are not determined by the lower levels, but only that they are not uniquely so determined.

The fact that organisms and machines exhibit properties that are neither characteristic of their individual constituents nor predictable from a knowledge of these parts conceived in isolation raises the problem of emergence: How are we to account for the occurrence of properties on higher levels of organization when these properties have never appeared on a lower level?

In a general sense, emergence signifies a class of some of the most familiar facts of the universe. The sound of two (or more) hands clapping, the texture of a woven fabric, the functional properties of a wire sieve, a geometrical pattern formed out of pebbles—all of these can be considered "emergents" in the sense that they represent higher levels of organization of their elements and might not have been predicted on the basis of a familiarity merely with these elements. Another type of example is provided by chemical composition: the properties of water or salt, so it is sometimes argued, could not have been predicted from a knowledge of the properties of their constituents taken separately. Applied to the case of living organisms, it could then be maintained that even if we possessed complete knowledge of the physical and chemical properties of every cell, including that of every molecule composing these cells, we would be unable to predict all of the properties of the total organism, or even of its internal physiology.

Thus construed, emergence presents no significant philosophical problem, however—at least not for the understanding of biological systems. There is nothing mysterious about unpredictability of this sort, and it could easily be eliminated if we included among the properties of the constituents certain of their relational or combining properties as well. Furthermore, it is not strictly true that we can never predict the "emergent" properties of new chemical compounds, since we do have a basis for predicting the color of organic compounds, for example. In fact, the microscopic and submicroscopic constituents of matter have in general become known to us only as a result of our investigation of structured entities which they have combined to form. Much of our knowledge of how the simpler entities behave in isolation derives from our observations of their behavior in combinations. And even in the cases in which knowledge of the elements precedes knowledge of their ordered aggregates, as in artifacts, for example, the properties that are said to "emerge" as a result of the act of combination can be seen simply as the realization of the potential of the individual parts. It is thus that wood has the potential of being fashioned into a desk or a house or a bowl, whose peculiar properties might be construed as the actualization of what was potential in the original lumber itself.

Emergence presents a special problem for biology only if the relationship between biological and physical or chemical properties is fundamentally different from that between physical and chemical properties. We know that processes occur in nature whereby higher degrees of organization are produced out of elements at a lower degree of organization, such as the formation of atoms from elementary particles and the formation of molecules from atoms: astrophysicists, for example, have exhibited a scheme which shows how protons are transformed into carbon, and chemists have shown that if elements (or atoms) are surrounded with an appropriate environment, the atoms are transformed into more or less complex molecules (41, p. 61). Emergence *must* occur, if the objects of any science represent aggregates of the objects of any other science; it is presupposed by the conception of a hierarchical structure of science. As far as biology is concerned, the problem lies not with emergence in general but rather whether biological emergence (i.e. "the emergence of life") is different from any other kind, whether or not it is possible to go from molecules to living systems in the same way that it is possible to go from elementary particles to molecules.

Emergence as a uniquely or characteristically biological phenomenon has often been associated with vitalism, the doctrine that living things are distinguished from nonliving things by virtue of a special life-force or vital principle that is present within them. As has already been remarked, this position is rejected by virtually all contemporary biologists and hence is not to be taken seriously as a presupposition of biological research. There is another interpretation of biological emergence, however, that is fully consistent with regarding organisms as physical mechanisms. This is the view of Polanyi, that because mechanisms, whether man-made or biological, constitute boundary conditions which are unspecifiable in terms of particulars belonging to lower levels, and which are governed by operational principles not capable of formulation in terms of physics or chemistry, they represent higher levels of organization that can have come into existence only through a process of emergence (*107,* pp. 382-404; *108,* pp. 29-52). There is thus a fundamental discontinuity between machines and living things, on the one hand, and inanimate nature, on the other. Emergence for Polanyi is associated not with the fact that the combining of elements may produce features which have not been observed in the separate parts, but rather with the fact that machines and living mechanisms have a structure or morphology that cannot be derived from physics and chemistry.

As we have already seen, the feature of any mechanism that distinguishes it from a nonmechanism is that the range of possible structures that constitute mechanisms of that operational type constructible from a given set of elements is a great deal narrower than the range of possible configurations of those elements consistent with the laws of chemistry and physics. Expressed in more technical terms, mechanisms such as clocks and living systems are nonholonomic: there are more dimensions in the space which can describe possible systems than there are dimensions to the actual development of a given system (see *155*, 1:219-20). It is essential to almost all man-made machines that they introduce nonholonomic constraints, such as ratchets, relays, switches, and escapements; these are the features by virtue of which the machine functions according to some human design. In biological systems these constraints consist in the hereditary mechanism, and it is an essential feature of this mechanism, if it is to function as a nonholonomic constraint, that a number of different configurations (such as DNA sequences) be equivalent energetically, that is, from the standpoint of physics. Unlike geological formations and ordinary chemical molecules, which are uniquely, or nearly uniquely, determined by the features of the combining elements and the thermodynamic characteristics of the resulting equilibrium, machines and DNA molecules necessarily exhibit the feature of appearing to have been selected from a number of physicochemically equivalent possibilities.

Whether or not the properties of those machines and living systems that depend on hereditary or nonholonomic constraints are çalled "emergent," there is clearly nothing mysterious or inexplicable about them, provided there is not a problem about how these systems came into existence. If one were to ask what it would be like for an emergent property not to be intelligibly emergent, we should have to reply that the occurrence of that property would have to be incompatible with the laws governing objects on the lower level. But if substances A and B were found to combine to produce a compound C with unanticipated or surprising properties, either this would constitute evidence of previously unrecognized properties of A and B, or, if the result were in explicit contradiction of predictions according to existing theory, then the result might indicate a possible refutation of that theory or of some part of it. Thus interpreted, the result might be unexplained, but it could not be regarded as unintelligible in principle. If, on the other hand, the behavior of the compound were found to be irregular and nonreproduceable, neither would this establish emergence, for an emergent character, as a scientific phenomenon, must be assumed to occur with regularity and to be capable of being fit into a consistent pattern. If a soul must be added to a machine in order to make it come to life, then life cannot be an emergent property in anything but a vitalistic sense.

A property is either ultimately intelligible within the framework of the laws and theories concerning the level from which it may be supposed to have emerged, or it is not. If it is so intelligible, there is little if any point to calling it *emergent,* but if it is not, then it is no more to be considered emergent than would a miracle that occurred as a result of divine intervention. Emergence turns out to signify either that which is predictable in principle (at least on a statistical basis) or that which could never come within the scope of any scientific theory.

Biology and the Origins of Biological Objects

The fact that a machine owes its particular shape to

having been constructed artificially according to operational principles that define it does not, we have seen, upset the thesis that machines are ultimately explicable in physicochemical terms. The same sort of point applies to biological systems, assuming they are to be regarded as physical mechanisms: the explicability of processes occurring within living systems does not depend on knowledge of how these systems came into being. On the other hand, the question of origins is a problem with which biologists have long been concerned, and biology, at least since Darwin, has included the study of how existing structures have originated. Although much of what comes under the rubric of biological research, including a large portion of anatomy, physiology, and biochemistry, has been carried out essentially independently of any assumptions that would deny that current species have all existed from eternity, there is a great deal of what goes on in biology for which this is not true. Not only is the origin of species a biological problem, but so are the origins of organs, physiological mechanisms, and biochemical pathways. There are many who insist that biology is essentially a historical discipline, and would affirm the centrality of evolutionary accounts of biological phenomena. Regardless of which aspects of biological inquiry are to be judged central, however, it can be said that the truth of some sort of evolutionary theory is relevant to, if not actually presupposed in, *all* areas of biological knowledge. All of biology has evolutionary implications, and these implications are part of biology.

The reason biologists have assumed the responsibility of accounting for the origins of the systems they study would appear to be that they are prepared to accept neither a biblical nor an Aristotelian account of the origin of species. Biology could very well exist without a historical dimension if it were assumed either that existing species have all existed from eternity or that they were all created by God at some point in time. Since biologists, unlike engineers, are ordinarily unwilling to assume that the systems they study are the products of a designing mind, they are faced with the burden of explaining not only the empirical possibility of living organisms but also the fact of their existence.

There is thus a definite logical connection between the fact that biology embraces evolutionary explanations and biologists' implicit denial that the systems they study may have been artificially constructed. The assumption that underlies the biologist's investigation of those physical mechanisms called organisms is that the history of these systems is a record which is explicable in terms of physicochemical theory when supplemented by data concerning astronomical, geological, and meteorological conditions. New species are to be explained as arising out of earlier ones in the same way that mountains and lakes are explained as having arisen out of earlier geological formations.

A theory of biological evolution can provide an answer to only part of the problem of accounting for origins, however, if evolution is interpreted to begin with the first organism. Biologists ordinarily are concerned only with living systems, but if the simplest organisms are taken as given, there still remains the fundamental problem of explaining how *these* physical mechanisms came into existence. It is at this point that the biologist could conceivably become uncomfortable. Whether or not abiogenesis, the spontaneous and natural formation of living systems from inanimate materials alone, is explicable in terms of physical theory, it clearly cannot be explained in the terms of biology. The biologist may be able to explain how organisms work, and how they arise from and give rise to other organisms, and he may even be capable of showing how an organism could be synthesized from its constituents, but he is not prepared to show how these parts could have come together *by themselves,* any more than an engineer could be expected to show how a steel suspension bridge could have come into existence purely as a result of natural forces without the intervention of an active intellect. If the biologist follows the problem of origins to its logical outcome, either he must assume that what he is dealing with are naturally occurring machines whose origins are in doubt, or he must assume that the task of providing a naturalistic account of the origins of objects of his domain can be achieved by the sciences outside of biology. The biologist, even the evolutionary biologist, must take a certain degree of organization as given.

From the standpoint of molecular biology, the problem of the origin of life is essentially that of explaining the self-generation of self-replicating molecules. It is the problem of spontaneous generation, expressed biochemically. If life was not created supernaturally, and if it did not simply develop from preexistent "seeds" present from the creation of the universe (whenever that was), life must have come forth from nonliving matter. Furthermore, this spontaneous generation must be conceived as having occurred at a unique or nearly unique time, or else the evidence upon which the theory of evolution is based, including the presence of the same kind of complex hereditary mechanism in both kingdoms of the living world, would not be explicable. As Blum has pointed out, if spontaneous generation has occurred with any degree of frequency, the whole evolutionary picture would continually be repeating itself, and it would in fact be extremely difficult to find any evidence of evolution at all (*14*, p. 173). If living matter actually derives from nonliving matter, then at some time in the history of the universe relatively simple substances must have come together to form self-replicating, self-modifying systems.

The prevailing view among astrophysicists, geologists, and biochemists interested in the problem of the origin of life is that present biological molecules are derived from molecules formed at a time when the earth's atmosphere contained no free oxygen (*11, 41, 101*). Preorganismal molecules may have been formed from simple molecules such as water, methane, and ammonia, and it has been shown in the laboratory that the basic biochemical molecules, including the constituents of nucleic acids and proteins, can be produced by electrical discharges. These molecules are presumed to have formed the constituents of a "primitive soup," which also came to contain more complicated molecules in the form of polymers, formed by the stringing together in linear order of similar or identical monomers or submolecules. It would have to have been at this stage that there appeared the mechanism of reproduction and replication of long-chain molecules. The final stage in the origin of life on earth would then consist of the formation of the simplest organisms through biochemical and structural transformations of these elements, leading to the development of internally organized systems.

Current discussion of the problem of the origin of life is concerned with working out the particulars of these processes and in showing them to be plausible. Although the greater portion of these investigations are theoretical, there is also a certain amount of experimentation with model systems being carried out. One aspect of the problem that has been receiving considerable attention is that of the evolution of the genetic code. From a mechanistic point of view, it is the possession of a hereditary mechanism, more than anything else, that distinguishes living from nonliving systems, and this mechanism requires a code in order to carry out its heterocatalytic function. Every organism contains not one but two types of polymers, proteins and nucleic acids, one of which catalyzes the synthesis of the other according to a universal code. How this code could have developed at all has represented a challenge that a number of molecular biologists have attempted to meet.*

The origin of living systems, as it is seen by those attempting to elucidate it, is a process of growth and self-complication of configurations of matter somewhat akin to the processes of biological evolution. Biological molecules, and biological structures, are understood as spontaneously generated structures that have survived. Molecules are assumed to have been formed randomly, followed by a chemical kind of selection. Furthermore, calculations based on astronomical data have been made to suggest that the universe contains at least 100,000,000 planets which can support cellular life as we know it on the surface of the earth (*18*). If the evidence that is cited for the development of the biosynthetic apparatus from nonbiological materials is accepted, then it is reasonable to believe that the same sequence of events as we can suppose happened here can have happened a great number of times elsewhere.

On the other hand, many scientists who have investigated the problem of the origin of life are still skeptical

*See, e.g., Woese *164*, chap. 7; *165;* Crick *27;* Orgel *102.*

as to the plausibility of any of the schemes that have been put forward. Some have challenged the idea that self-replicating systems could ever have arisen at all purely on the basis of chance occurrences, and have suggested that some as yet undiscovered principles may in fact be operating. One of the problems is that evolution, whether chemical or organic, seems to produce partial results that represent recognizable progress toward a goal, results that play the role of a stable subassembly, like a defective lock on a safe whose dials click at the proper settings (*126*, p. 96). Thus Mora has pointed out that the principle of selection, in the sense in which it applies to chemical molecules—for example, in describing selective interaction between enzyme and substrate or antigen and antibody—cannot be used to account for the building up of the more improbable and complex from the more probable and less complex, without additional assumptions such as that stability increases as complexity increases (*41*, pp. 47-48). It has even been argued that, according to standard quantum mechanical theory, the probability that there are self-reproducing structures is zero, and moreover, that it is impossible in principle to calculate the behavior of biological units on the basis of the laws of physics (*162*). A living system, furthermore, requires *two* kinds of self-reproducing informational macromolecules, each of which seems to require the existence of the other for its synthesis. What needs to be accounted for is the appearance not merely of a persistently self-reproducing unit but of a self-reproducing unit the order of whose subunits determines the order of subunits in polymers of the other type. Not only have the details of the self-generation of the molecular replicative mechanism not been worked out, but it is commonly conceded (see, e.g., *11*, p. 145) that no one has put forward even a plausible account of how such a process could have occurred.

To the extent that the problem of the origin of life is a scientific one, it is not going to be solved by philosophical reflection. It may be of philosophical interest, nevertheless, to consider some of the presuppositions of the various positions taken by those who have addressed themselves to the problem. For example, the general scheme of the history of the origin of life on earth has been defended as a "reasonable, continuous, hypothetical story, a myth" (*11*, p. 35). The assumptions are that life *did* evolve, that the essential mechanisms did appear at some stage, and that we may treat the problem of working out the stages as essentially a puzzle to be solved. On the other side, there are those who are less sanguine about explaining the problem with existing knowledge and who have warned that this attitude can lead to an avoidance of the complexities of describing biological relationships (see, e.g., Mora, in *41*, p. 50). Scientists can become so impressed by the possibilities of building up a hypothetical historical picture of the origins of life at all that they may be prone to ignore or to underemphasize logical difficulties and matters of detail.

Arguments on both sides have often been characterized by somewhat less rigor than scientists have usually demanded of themselves in less speculative areas. For example, a number of people have tried to extend the biological notion of natural selection to account for the building up of the first self-replicating system, forgetting that the Darwinian conception already presupposes the existence of a mutable, self-replicating system. What survive in biological evolution are self-reproducing types, not stable configurations of matter. On the other hand, there have been probability arguments put forward that purport to show that abiogenesis is itself a physical impossibility, arguments that call to mind "proofs" that the bumblebee cannot posibly fly. A priori arguments based on improbability apply not to the genesis of life itself but to the adequacy of a particular hypothesis made to explain it. When we say that there is little or no chance that something will happen (or has happened), we are implicitly assuming that the mechanisms we are considering are the only ones possible.

Current interest among biologists in the problem of the origin of life, though it may be viewed as essentially an extrabiological issue, represents an affirmation that biology is a basic science which operates on a variety of levels, and a denial that it is essentially akin to engineering, amounting to the study of the working of certain machines. Commitment to the view that life has its origins in inanimate matter is of a piece with belief in the unity of science. The origin of life by natural causes is generally regarded as a reasonable scientific hypothesis, but its denial is not incompatible with anything that falls under the rubric of biological research. It is in this sense a metabiological thesis, a thesis about biology itself. The fact that it is subscribed to by a majority of contemporary biologists is indicative of their conception of the position of biology among the sciences.

Biology and the Use of Paradigms

Biology's subject matter thus consists of a class of naturally occurring machines, machines that are assumed to be descended from primitive protomachines, the original progenitor of which was self-assembled. Biology is concerned with explaining the behavior of these machines and their internal parts. It was promised in chapter 2 that in this chapter we would consider what determines the adequacy of explanation in biology apart from formal logical structure. Any explanation, in order to be effective, must presuppose the intelligibility of certain occurrences or states of affairs in terms of which other phenomena are explained. These fundamental intelligible types—or "ideals of natural order," as Toulmin calls them (*148*)—constitute the norms to which those happenings in the world that do require explanation are referred. To understand what makes a scientific explanation intelligible is to recognize the paradigms and explanatory ideals of that science.

In physics, or at any rate in classical physics, the most general paradigm is material bodies moving in accordance with Newton's laws. Straight-line motion at a uniform speed is taken as self-explanatory, and all other forms of motion are to be analyzed as departures from it. These ideas are embodied in what Harré (*55*) has called the "corpuscularian philosophy," according to which the world is conceived as made up of individuals whose basic properties are mass, extension, and shape, and whose fundamental interactions are impact and gravitation. Explanations in chemistry and in later physics can be seen as employing an extended version of this paradigm, enriched by the inclusion of electrical properties. Chemical change thus came to be conceived as explainable in terms of the rearrangement of atoms, and nuclear physics could represent atoms as miniature solar systems.

To the extent that biology consists of biochemistry and biophysics, its paradigms are the same as those of the physical sciences. To understand a life process, for anyone committed to the use of this paradigm, is to elucidate it in molecular terms, to analyze it as a special case of ordinary chemical transformation. The Watson-Crick model of the replication of genetic material represents a classic instance of this. Biologists employ other paradigms as well, however. One of these takes the normally functioning organism as itself an ideal of natural order. In pathology this serves as the norm to which anything that is identified as an irregularity is contrasted; in anatomy and physiology this paradigm provides the basis for the identification and description of structures and processes in terms of their contributions to the functioning of the organism as a whole. Research in brain physiology, for example, is commonly directed toward elucidating the functional organization of the brain. The discovery that a structure controls a certain piece of behavior might thus constitute the terminus of a particular line of investigation.

Another type of paradigm often assumed by biologists is a historical one. Thus it is sometimes maintained that biological phenomena can be understood only if one knows how they came into existence (see, e.g., *131,* p. 105; *101,* pp. 30-37). Or it may be the normal development of things through their life cycle that is taken as normal and natural, not in need of explanation. For the first case an evolutionary account of some adaptive characteristic would exemplify the explanatory ideal; in the second case a developmental paradigm would provide the background for the ontogenetic explanation of such an abnormality as a harelip. The biologist who expects an explanatory account to answer the question, "How come?" has adopted a historical paradigm.

It should be clear that the assumption of any of these explanatory paradigms or ideals of natural order is neither true or false in any straightforward sense. The difference between them consists in which states of affairs are selected as "natural" or "normal." Furthermore, as we have seen in chapter 2, these different types of account need not be mutually incompatible. Historical accounts in biology, for example, are generally assumed to be consistent with physicochemical explanations, as are functional analyses and macrolevel physiological explanations. There can be said to operate what amounts to a hierarchy of paradigms, in which higher-level paradigms are explicable in terms of lower-level ones. Biology employs both same-level explanations, in which phenomena are explained by comparing them to other, less puzzling occurrences of the same general type, and reductive explanations, in which phenomena—often the same phenomena—are exhibited as complex instances of self-explanatory happenings on a lower level. The choice is essentially a pragmatic one.

Scientific controversies can often instructively be viewed as clashes between competing paradigms. The difference between the mechanistic and the organismic viewpoints in biology, for example, may be seen as a difference in paradigms, the difference between seeking explanations in terms of physicochemical principles and attempting to explain biological processes in terms of the system as a whole. One attitude is that life processes are understandable when they can be analyzed in physicochemical terms; the other is that biochemical processes can be made sense of only in terms of their roles within living systems. It is the same difference in attitudes that is revealed in Barry Commoner's remark (cited in *129,* p. 14) to the effect that life is the secret of DNA, rather than the other way around.

The question of biology's autonomy, discussed in

chapter 1, is also related to the matter of paradigms. The person who stresses the autonomy of biological accounts is one who finds or expects to find acceptable explanations in terms of distinctively biological categories. The direction of biological research that accords with this attitude is toward achieving explanations of biological phenomena in terms of increasingly general principles of biological organization. The explanatory ideal that is relevant here appears to be that of a hierarchically organized machine. A theory like von Bertalanffy's general system theory would amount to a working out of that paradigm. There are in fact very few biologists today who are attempting to develop this paradigm or to provide explanations of biological phenomena by subsuming them under the sorts of principles that such a theory seeks to put forward. Whether this indicates that modern biology is still at a primitive stage, or that it is well along the "right" methodological track, is a reflection of the fundamental difference in outlook of the autonomist and the antiautonomist in biology.

Presuppositions of Biological Taxonomy

Contemporary biology, whatever the level of investigation, is consistently naturalistic: it assumes that biological phenomena need no supernatural explanation, that they are governed by the same principles as obtain throughout the rest of nature. It is not the naturalistic paradigm that gives biology its distinctive character, however, nor is it the use of such general paradigms as embodied in the mechanistic or holistic modes of explanation. Biology's peculiar nature can only be revealed in its internal or subsidiary paradigms and in the assumptions which their use embodies. Explanations within the various areas of biology are found acceptable in proportion to the acceptability of the general principles that they presuppose, and these principles may in many cases be themselves biological.

The most obvious and universal of biology's conceptions that provides a background for description and explanation in biology is the theory of evolution, and one of the aspects of biological description that most clearly reflects this background is that of the classification of organisms. An essential requirement for the success of any science is that it employ a set of categories that are both convenient and fruitful. Biology's taxonomic categories represent a sorting of that science's primary objects into classes according to properties to which biologists have attributed importance in attempting to characterize the domain of living phenomena. The theory of evolution provides a framework for such sorting.

Prior to the rise of evolutionary theory, there was no independent theoretical basis for cataloguing biological specimens. Classification consisted in arranging plants and animals into groups according to their appearances, their habits, and their relations to each other. A taxonomic system would be considered effective if it managed to lump together organisms with many attributes in common, so that one set of diagnostic characters-in-common could be used as reliable indices for a number of other characters. Linnaeus and others who conceived of natural history as the search for the "natural" method of classifying organisms believed themselves to be out to discover the pattern according to which God had ordered the universe. Their only presupposition was that nature represents a rationally ordained system of means and ends.

The theory of evolution not only offered an explanation of the patterns discerned by earlier taxonomists, but also contributed a principle of classification: organisms were to be catalogued according to phylogenetic descent. As it turned out, classification carried out on this basis has preserved many of the Linnaean groupings, in spite of the radical difference in conceptions of how these groups came into being. What seemed a rational and convenient system of naming and classifying also served as a way of representing the sequence of events that constitute evolutionary history.

On the other hand, the use of evolution as a basis for taxonomy raises certain problems that would not have come up in a nonevolutionary context. If species are assumed to be fixed, then it is reasonable to expect there to exist clear lines of demarcation which separate them from each other. To say that species are "real," as opposed to "artificial," is to say that a taxonomic group is defined by the fact that its members all share certain essential characteristics. Evolutionary theory, however, suggests that there are no such groups, and that there is no basis in nature for distinguishing essential and accidental characteristics in organisms. A taxonomic system based on phylogeny would seem to preclude the possibility of natural kinds altogether.

It deserves to be pointed out that the preevolutionary notion of a natural method of classification in biology was itself controversial. Thus Linnaeus's distinguished contemporary Buffon, as Greene points out (*50,* p. 143), considered the dominant characteristic of nature to be an endless proliferation of forms, each form grading imperceptibly into others. Systems of classification were sometimes viewed merely as attempts by man to impose his own conceptualizations upon nature. In the absence of a theory of the origins of variation it was impossible to justify the procedure of stressing similarities and ignoring differences that the implementation of a strict classificatory policy demanded. Part of the achievement of evolutionary theory consists in the fact that it offers an explanation both of the occurrence of groups of organisms sharing a number of common attributes and of the failure of nature to exhibit totally discrete types. Systematics based on descent and variation account both for the discontinuities and for the continuities that are observed within the range of terrestrial organisms. All of this follows from making biological (i.e. genetic) relationships, rather than morphological similarities, the fundamental principle.

The most salient feature of a taxonomic system based on phylogeny is that it treats the population, not the individual, as the basic taxonomic unit. Since the members of any species differ from one another in various ways, and since there is no feature or set of features that is both necessary and sufficient for membership within any species, there can be no individual specimen which serves to define the category. The class of organisms that fall within the range of any single classificatory term are bound together not by the possession of certain essential characters but by what Wittgenstein called "family resemblances," "a complicated network of similarities overlapping and criss-crossing."* It follows that a precise definition of any species category is in principle impossible, and that the range of a particular species-term can be understood only on the basis of knowing what characters are widely distributed, though not necessarily universal, among members of that species. In

*163, Pt. I, secs. 66-67. This notion was first applied to biological taxonomy by Beckner; see 9, pp. 22-23.

other words, to recognize a specimen as typical presupposes knowledge of a population and the characters which members of that population typically possess.

What ties the members of a population together for the evolutionary taxonomist are their genetic connections, their relations of biological kinship. It is for this reason that interfertility is taken as the single most important criterion for placing individuals within a single species-category: species are defined, albeit vaguely as potentially interbreeding groups of organisms. Morphological similarities serve as effective indices of phylogenetic relationships, and these in turn suggest considering other possible bases for comparison, such as properties of a biochemical or an ethological nature. Evolutionary taxonomy is possible because individuals with similar ancestry have similar observed characters. It is not the possession of these characters that determines the species of an individual, however, but rather the position on the phylogenetic tree of the population in which it occurs.*

Presuppositions of Modern Evolutionary Theory

The fact that the choice of biology's taxonomic categories is determined by a theory of evolution does not imply acceptance of any particular account of the mechanism of evolution, however (unless what we *mean* by "evolution" is "evolution by natural selection"). The hypothesis that all existing forms are descended from a much smaller number of ancestral forms can provide a

*Recent discussions of the species concept include those by Beckner (*9*, chap. 4), Lehman (*77*) (see also reply by Munson [*98*] and Lehman's counterreply [*80*]); and Ruse (*114*), with reply by Hull (*64*). For further references, see Hull 65.

basis for taxonomic classification without presupposing any specific mechanism of evolutionary change. As we have seen, accepting any evolutionary theory at all involves making certain assumptions about the nature of living things. A number of additional assumptions are associated with endorsing any particular theory of the nature of the process, such as the currently prevailing neo-Darwinian theory of evolution by random variation and natural selection. As a means of examining this theory's presuppositions, it will be useful to consider it in the context of some of the objections that have been made to it.

According to the neo-Darwinian interpretation of biological evolution, nothing but random variation and natural selection is needed to account for the proliferation of species which inhabit the earth. Mutations are understood as essentially chance phenomena, and it is the selection process alone that is credited with producing all that is systematic and orderly in the biological sphere. Furthermore, it is assumed that every evolutionary change contributes to the survival or reproductive potential of the organism that undergoes it. In other words, all change is adaptive; otherwise it could not have been preserved.

Although this is clearly the dominant view held by biologists, it is by no means universally accepted, at least not in its more austere formulation. There have been a number of challenges to the neo-Darwinian interpretation which have raised doubts as to whether all of the known results of the evolutionary process would be possible on strictly Darwinian or neo-Darwinian assumptions. What Darwin demonstrated was that, given multiplication, heredity, and variation, natural selection follows; he did *not* show (nor did he believe) that *all* biological features can be accounted for by this mechanism. The neo-Darwinians have tried to show that they can. The issue is thus not over whether evolution by natural selection occurs, but rather whether this is the only mechanism that needs to be supposed.

The problem comes more clearly into focus when it is pointed out that natural selection itself has contributed nothing to the actual result of evolution, other than restricting the number of forms which have survived. In other words, if no selection had occurred at all, all of the known forms of life would have appeared in any case, along with a vast number of others. There is nothing in the neo-Darwinian theory that suggests that the frequency of selectively advantageous mutations is increased by elimination of the unfit. It follows that the principle of natural selection is of no use whatever in explaining the evolution of the high degree of complexity and functional organization that characterize living things.

Recognition of the fact that natural selection is only the editor and not the author of evolutionary change has contributed to an attitude of skepticism on the part of many with respect to accepting random search as an adequate mechanism for producing such change. Waddington, for example, noting that the number of protein molecules that have ever existed is but a minute fraction of the number that are possible (10^{52} as opposed to 10^{325}), and that evolutionary transformation requires that all intermediate stages be viable as well, has suggested that there may be restrictions affecting the "randomness" of mutation, and that phylogenetic change may be governed by higher-level principles which lead to a channeling or "canalization" of these processes (*155*, 1:111ff.). The problem to which these people usually point is that evolutionary changes, especially in higher organisms, generally involve more than one single-gene mutation and that most such changes are harmful. Since almost all random changes of any message, whether embodied in a DNA sequence or in a natural language, render the text meaningless, some have considered it necessary to assume something more to exist in biological systems in order to account for the apparent directionality of evolutionary change.

Another type of challenge to the orthodox neo-Darwinian interpretation of evolution has to do with the requirement that all evolutionary change be selectively advantageous. One problem in this area that has not been fully resolved is that of explicating the mechanism that stops some mammals from multiplying beyond certain limits before they have destroyed all of their food resources, the difficulty being that of explaining how a group could pass on a reduced propensity to mate (see *155*, 2:83-95). There is ample evidence that such self-limitation of animal population does occur; what is unclear is whether and how neo-Darwinian selection theory can account for it, since what seems to occur is that the variants which come to prevail will be those that leave *fewer* offspring than those which are supposedly being "phased out."

A development that has been thought to challenge the Darwinian or neo-Darwinian account of selection still more directly has been the presentation of evidence that suggests that most evolutionary changes on the molecular level are selectively neutral (*73*). Not only are there large numbers of DNA base-pair changes which, owing to the "degeneracy" of the genetic code (the fact that more than one codon or three-base sequence may code for a given amino acid), have no effect on protein structure ("synonomous mutations"), but it has been shown that many of those changes in DNA which *do* result in altered proteins nevertheless are fully equivalent to the original form with respect to natural selection. It has thus been estimated, for example, that there are literally millions of hemoglobin variants within the total existing human population. Natural selection can be assumed to operate on morphological, functional, and behavioral levels without implying that all evolutionary changes on the molecular level must have an effect on the fitness of the organism.

The occurrence of selectively neutral change on the molecular level need not be taken as inconsistent with a neo-Darwinian interpretation, however, if the latter is recognized as applying only on the phenotypic level. Furthermore, the hypothesis of the fixation of selectively neutral mutations implies a much higher mutation rate than can be determined on the basis of observable effects on function and hence provides a possible explanation of the finding that evolution seems to have proceeded at a greater rate than natural selection may be able to account for, given the fact that many, if not most, evolutionary changes involve more than one amino-acid substitution. What the idea of selectively neutral mutations does show, on the other hand, is that evolutionary change cannot be understood entirely in terms of classic evolutionary theory.

The most radical of the challenges to the neo-Darwinian interpretation of evolution are those that question the entire mechanistic framework of modern evolutionary theory. The thesis that evolution is the result of the joint action of random variation and environmental pressure is thought to constitute an illegitimate reduction of life to an essentially Cartesian and lifeless world whose only fundamental principles are chance and necessity. This is what Marjorie Grene characterizes as "the faith of Darwinism," the belief that nature is to be construed as "a mechanically interacting aggregate of machines," (*51*, pp. 185-201). Such a conception is held to be inadequate because it fails to take cognizance of the harmony of adaptations and the comprehensive ordering of biological forms. What is thought to be called for is "a richer metaphysical vision," one which acknowledges the multiplicity of forms of beings and the possibility of preexisting types and inherently teleogical or self-fulfilling processes.

The fact that very few if any contemporary biologists take seriously the alternatives suggested by this interpretation is evidence of the pervasiveness of the Cartesian attitude. Biologists *are* committed to attempting to explain the phenomena of life in terms of the nonliving, and evolutionary phenomena are no exception. The property of being alive or being associated with a living system is simply not treated as a logical primitive within biological science. Whatever difficulties evolutionary biologists face in accounting for biological structure or form and the apparent goal-directedness of evolutionary change, it is nearly universally assumed that the resolution will arise out of neo-Darwinian theory and its extensions, and not out of considering "phylogeny as an ontogeny writ large . . . the history of groups as expressing a fundamental rhythm still, in its intimacy, unknown to us, but analogous to the rhythm of individual development" (*51*, p. 194). Whether or not such an alternative conception constitutes an ultimately superior interpretation of the facts of evolution, as some philosophers and theologians have maintained, it is clearly not one that is taken seriously by the life sciences as they exist today.

The differences in various attitudes toward biological evolution can be seen by comparing the paradigms assumed by each interpretation. For the orthodox neo-Darwinian the fundamental intelligible type is exemplified by the case of industrial melanism, in which it was shown that the black mutant of the common light-colored moth occurs with far greater frequency in industrial areas, where it is less exposed to predators, than it does in rural areas, the two varieties differing in a single Mendelian gene. The paradigm for a mechanistic but not strictly Darwinian account would be that of the spontaneous formation of crystals within a solution, or perhaps the construction of an engineered structure such as a bridge according to rational principles out of randomly produced components. The self-explanatory type for evolutionary explanation associated with the conception of nature as a growing organic whole would be essentially that of a growing plant or of an animal behaving purposively. The aim of most biologists is to achieve explanations of the first type, although many are prepared to accept some of the second as well. Virtually none would consider appealing to the third as an explanatory model, for the biologist, almost by definition, can be expected to begin where one of these accounts would leave off.

Presuppositions of Functional Analysis

The effective dominance of Darwinian thinking in biology may be revealed in one of biology's most characteristic underlying assumptions, namely, that every structure, every process, every piece of behavior associated with an organism has a function. Expressed in traditional terminology, it is assumed that every event within the domain of biological science has a final cause. This principle, which plays a heuristic role and guides investigation in a number of areas within biology, including anatomy, physiology, and ethology, is a direct consequence of the neo-Darwinian theory, which requires that all evolutionary change be adaptive.

The proposition that every biological substructure has a function, though implied by the neo-Darwinian theory of evolution, does not itself imply that organisms must have evolved, and in point of fact the concept of biological function is older than the theory of evolution. The notion of function is compatible not only with alternative interpretations of evolution, such as Lamarckism and autogenesis (the view that evolution is essentially the unfolding of what was there from the beginning), but also with the view that species are the results of independent acts of creation. The important question to arise is whether functional accounts presuppose some sort of design-principle or other.

According to the view that the notion of function is essentially linked to that of design, functional accounts are parasitic on purposive explanations. The reason why it is not ordinarily considered satisfactory to say that the function of water's expanding upon freezing is to make underwater life possible (even though it does have this effect) is, on this interpretation, that we do not conceive of the world as having been designed by a Creator. To ask why something is the case and to be prepared to accept a reply in terms of its function, so it is argued, is to assume that the phenomenon in question, if not actually the result of someone's intentions, is explicable in terms which indicate why it so appears.*

If functional analysis is understood simply as revealing the mutual adaptedness of parts within a system, on the other hand, no connection between function and purpose or intention need be assumed. Functional organization is epistemologically prior to intention in the sense that one infers an antecedent design or purpose *from* observations of functional interrelatedness, not the other way around. While it is true that every book presupposes an author, it is not necessary to assume anything concerning the nature of its authorship in order to be able to identify it as a book; the same applies to the identification of functions. As was pointed out in chapter 2, the connection between serving a function and having been created for a purpose is a contingent one. Biology employs functional explanations by virtue of the assumption that the parts and processes of living things contribute to the survival and maintenance of the system as a whole, not because it is assumed that these elements were intentionally created to do so. The very fact that evolutionary biologists have been able to continue using the traditional concept of function without having to change it strongly suggests this interpretation. What has changed is not the concept of function but rather the type of explanation that is given to account for its applicability.

It follows that the reason we do not ordinarily use functional concepts in dealing with such matters as the expansion of water upon freezing is simply that we are not accustomed to regarding inanimate nature as ex-

*Argument presented by Jerome Shaffer, in discussion.

hibiting a thoroughgoing mutual adaptedness of constituent parts. The prevailing conception of nature is not an Aristotelian or teleological one. On the other hand, it is interesting to note that "functional" accounts of this sort do contribute to our understanding of the world by exhibiting certain causal dependencies involving some of its gross features, even though we may not be committed to viewing nature as a system in which everything plays a functional role. Machines, organisms, and societies are our paradigms of functionally organized systems and constitute the domain in which the use of functional concepts is regarded as "natural." If the extension of these concepts to phenomena that fall outside this domain appears "unnatural" to us, it is because we do not conceive these phenomena as elements or aspects of a system in which everything has a function.

To ask why something is the case and to be prepared to accept a reply in terms of its function presupposes that it makes a contribution to a system in which it participates. The fact that the expansion of water upon freezing or the volatility of alcohol (which makes distillation possible) satisfies this precondition, but nevertheless fails to elicit the corresponding "Why" question, shows only that the contributory dimension of the phenomenon is not taken into active consideration. And the reason why it is not is a matter of ultimate presuppositions about the nature of the world—metaphysics, in other words.

Teleology and the Presuppositions of Ethology

Another area of biological science whose presuppositions are generally regarded as consequences of the theory of evolution is that of ethology, the study of animal behavior. Characteristic behavior patterns can be viewed as the results of natural selection in the same way that morphology is. Thus Darwin observed that "instincts are as important as corporeal structure for the welfare of each species" and that "under changed conditions of life, it is at least possible that slight modifications of instinct might be profitable to a species" (*29*, p. 209). The neo-Darwinian generalization of this suggestion is that every piece of purposive behavior is to be explained in terms of species survival. Ethology becomes essentially a ramification of functional analysis.

Like the principle that every biological structure below the level of a total organism has a function, however, the proposition that all behavior is functional serves more as an investigatory maxim than as substantive assertion. Each segment of behavior is approached with the question of how it contributes to the preservation and maintenance of the individual or species in which it occurs. Tinbergen, for example, writing about the social behavior of animals, asserts that when we speak of cooperation between individuals, we assume that it serves some end (*147*, p. 2). Cooperation between male and female enables them to come together for mating. Cooperation among members of a "family" or a group serves to protect the young and to prepare them for life as mature members of the group. Even fighting between animals of the same species is seen as useful; an example is territorial fighting, which has the effect of spacing out the members of a population over the territory available, and hence serves to ensure some sort of regulation of population density.

On the other hand, not all behavior can be linked to survival quite so easily or directly. In some cases apparently useless behavioral acts, such as a dog's scratching a hard floor before lying down to sleep, are explained as "behavioral rudiments," vestiges of formerly useful actions still performed by a species, analogous to adaptively useless morphological features often explained in terms of a species' imputed prehistory (*62*, p. 164). Other forms of behavior, such as displacement activity (defined as irrelevant acts resulting from a conflict of two drives), have been described without having been adequately accounted for on a functional level. Thus it has not been established conclusively whether such behavior as the sudden performance of feeding motions among fighting domestic cocks, or the casual brushing of the snout and whiskers by a tame rat exposed to an unfamiliar environment, is to be explained as an outlet through which strong but thwarted drives can express themselves in motion, or as the manifestation of a pre-existing third drive once the inhibiting effects of two stronger motivations have been mutually annulled (*5*, pp. 155-56; *62*, pp. 189-93). On either hypothesis, we lack an explanation in functional terms as to why the particular forms of displacement activities are performed.

Still another type of challenge to the assertion that all behavior is functional is presented by those activities in which the animal seems to be moved solely by curiosity or by an apparent drive simply to attain competence in the performance of a particular motor skill. For the former case, the attempt has been made to subsume a wide range of examples which include apparently random, unrewarded wandering, solving mechanical puzzles, and choosing to listen to changing patterns of sound, under the general heading of exploratory behavior (*5*, pp. 33-38), without offering any explanation as to why these particular forms are assumed. The latter sort of behavior, which also includes various games and other activities apparently done for their own sake, tend either to be left unexplained or to be "explained" in terms of such notions as Lorenz's "perfection-reinforcing mechanism" (*82,* p. 75)—which is another way of saying that they are done for their own sake. The claim that every piece of purposive behavior does something or other for the organism that carries it out can be made trivially and necessarily true by counting as a contribution anything that satisfies an apparent purpose, whatever it is, regardless of whether what is done contributes to the survival of either, the organism or its species. The principle that all behavior can be accounted for in functional terms has been an extremely fruitful one, but only because it has not been permitted to collapse into triviality: it is useful because it may be wrong.

The other important presupposition of a science of animal behavior based on a Darwinian or neo-Darwinian theory of evolution is that there are no such things as purposes, intentions, and final causes, and hence there is no need to invoke them in order to provide an explanation of the way animals behave. Purposive behavior is supposed to be explicable naturalistically, in terms of instinctive mechanisms and the concepts of stimulus and response, in the same way that evolutionary change is explained nonteleologically, as the result of random variation and natural selection. The differences that make the natural selection of behavioral variants possible are genetically determined; the behavior itself is dependent on both external and internal factors. What are inherited are specific neurological mechanisms which, when triggered by specific releasers or sign stimuli, give rise to motor activities that constitute the appropriate behavior pattern. Modern ethology consists in the objective causal analysis of instinctive behavior, and comprises investigations of a comparative, phylogenetic, ontogenetic, physiological, and biochemical nature.

In contrast to this way of understanding the behavior of animals, it has sometimes been objected that the ideal of objectivity breaks down at some point. Though it is acknowledged that much of instinctive behavior, such as nest-building, is determined by a series of rigid and precise mechanisms, it is argued that other activities described by ethologists (for example, the assuming of a threat posture by herring gulls) can be interpreted only in terms of human subjectivity. Understanding animal behavior, so it is argued (Polanyi *107,* p. 364), is possible only if we are able to imagine that we would do if placed in the animal's position. Marjorie Grene (*51,* pp. 213ff.), citing Tinbergen's investigations of the behavior of herring gulls, tries to show that this is what ethologists in fact do. The assuming of an upright threat posture, for example, he describes as "full of meaning" for all of the gulls; other situations are described as "really loaded with hostility." In addition, Grene notes that Tinbergen speaks of "genuine personal ties" among gulls, and that he admits having come to know the birds he was studying as personal acquaintances. Ethologists, she insists, observe individual living beings and must perforce deal with them as individual centers of activities. To describe animal behavior patterns is to refer to them "in terms of an understanding of the whole life of the organism or the community of organisms." To do so is to recognize them as, in some sense, persons, so it is concluded, and to acknowledge a subjective or unspecifiable component of the science.

Some might attempt to reply to this argument by insisting that ethology uses personal or anthropomorphic expressions merely in a metaphorical way, and that these could in principle be replaced by strictly behavioral locutions. Such a translation program, however, like all behaviorist reductions of mentalistic discourse, would be impossible. An alternative response, one which I think is correct, is to accept the appropriateness of the nonbehaviorist description but to deny that it implies that such knowledge is irreducibly personal or subjective. To conceive of a rat as a center of appetites and interests, so far as its behavior is concerned, is logically no different from conceiving of a television set as a center of sounds and pictures. There is nothing subjective about recognizing an individual as the logical subject of a behavioral or motivational state. Recognizing an individual living thing as a sort of "person" can be explicated as knowing what sorts of behavior to expect under given circumstances. Even Polanyi's suggestion, "that nothing at all could be known about an animal that would be of the slightest interest to physiology, and still less to psychology, except by . . . identifying ourselves with a centre of action in the animal" (*107*, p. 364), can be construed merely as calling attention to the point that there can be no understanding of bodily movements as components of purposive behavior unless they are viewed in the context of a whole series of movements and their ambient circumstances—in other words, the sorts of things that the organism itself is responding to, whether consciously or unconsciously. The explanation of animal behavior need be no less objective by virtue of being derived from a consideration of the animal's particular situation or point of view, so long as this is imputed solely on the basis of his observable behavior.*

*For a way of interpreting "subjective" or mentalistic language that does

Another type of attack on the mechanistic approach to animal behavior has been the attempt to show that it is impossible to account for any sort of purposive behavior at all in an entirely nonteleological manner. Charles Taylor has argued, for example, that it is useless to try to account in terms of stimulus and response for goal-directed activity such as the depressing of a lever by a rat in order to obtain a food pellet, since the limits of the response may be wider than those of a single motor habit: a rat who has been trained to depress a lever with his teeth will, if he is muzzled, use his paws to achieve the same end (*143*, p. 208). Furthermore, in the case of instinctive behavior such as the food-seeking behavior of animals that hunt, the "releasing stimulus" turns out to be a general type of situation in which a particular sort of motor activity is required for the attainment of a goal, rather than a specific "releaser"; the releasing situation, so it is maintained, cannot be defined except in teleological terms, namely, as one which requires a particular type of motor response (pp. 210-11). An ethological theory such as that developed by Lorenz and Tinbergen is held to be inadequate because it cannot delineate the internal conditions that facilitate appropriate behavior patterns without resorting to teleological forms of explanation, and because it has been unable to account nonteleologically for the variation within behavior patterns as a function of the situations in which they occur (pp. 229-33).

None of these arguments constitutes a fatal objection to the validity of the ethologists' enterprise, however. Concerning the first example, there is no reason not to

---

not commit its user to the existence of minds or other irreducibly non-physical entities, see the author's *128;* also Davidson *31.*

suppose that the training of a rat in bar-pressing in order to receive food involves higher-level neurological mechanisms which would make the use of an alternative motor mechanism simply a matter of course. Many ethologists in fact stress the role of hierarchical organization of the mechanisms that govern animal behavior. Nor should there be anything puzzling about the suggestion that the releasing stimulus be characterized as a "situation," so long as it is not denied that the situation presents certain cues to which the animal reacts. The problem does not differ in principle from that regarding an organism's apparent "generalization" to a specific stimulus-type from a single or limited number of stimulus-tokens. Finally, to say that a releasing situation can be defined only in teleological terms is to say merely that we need to resort to these concepts as a way of picking out those states of affairs that are causally relevant in the effecting of the appropriate action pattern. The same applies to the delineation of internal conditions (innate releasing mechanisms) and to the characterization of adaptively variable patterns of behavioral response. The use of teleological categories as means of lumping together the members of classes of internal and external states of affairs in no way constitutes a denial of sufficient causation between perceived situations, internal conditions, and patterns of resulting behavior.

Mechanism and teleology are not incompatible. A mechanical system, as Russell put it, is one which has a set of determinants which are purely material; a teleological system is one in which purposes are realized (*115,* p. 201). A mouse, real or mechanical, running a maze is an example of a teleological system. To describe what the mouse is doing in terms of trial and error is to employ a teleological mode of discourse. Each step, however, can presumably be explained mechanistically. Ethologists are concerned with systems that are describable in terms of the realization of purposes, and what they try to do is to provide mechanistic accounts of these systems. It is quite beside the point to insist that teleological accounts are useful and that they are needed for the identification on the macroscopic level of what is happening. The ethologist assumes only that the teleological account is not the only possible explanation.

Mechanistic explanations of purposive behavior are impossible only in the sense that they cannot do what teleological explanations can, namely, indicate that for the sake of which a piece of behavior is being performed. This difference is a feature only of the modes of discourse, the manner of construing the events under consideration, not of the states of affairs themselves. Since the language of purposes and intentions is not reducible to language of behavior and mechanical processes, there is clearly a sense in which purposive behavior cannot be explained mechanistically. But the ethologist is not concerned with explaining away purposiveness or with providing mechanistic or behavioristic translations of teleological descriptions; rather, he is concerned with determining the origin and functions of instinctive behavior patterns and with elucidating the physiological processes that mediate them. What ethology's mechanistic presuppositions rule out is the possibility that there is no causal chain to be discovered.

Biology and the Concept of Life

It is rarely mentioned that what is distinctive about the systems that biologists study is that they are alive. The reason, of course, is that life—or living things—is what biology is about. Biology exists because there is a radical difference between things that are alive and things that are dead, and because there seems to be a fundamental distinction between things that are either alive or dead and things that are neither alive nor dead. "What is life?" is the basic question that biological research is engaged in attempting to answer, even though the question itself is not ordinarily expressed.

We discern two separate tendencies among biologists with respect to the concept of life. These are revealed in the difference between speaking of the "characteristics" of life and speaking of the "building-blocks" of life. According to the former, life may be thought of as having a quality of its own, and its existence may even be treated as an elementary fact that cannot be explained but must be taken as a starting point in biology, analogous to the quantum of action taken as forming the foundation of atomic physics. The second conception views living aggregates of matter as essentially continuous with nonliving ones, and looks for ways of explaining the peculiar characteristics of living systems in terms of the same laws and principles as apply to matter in general. If life is to be defined or explained, as opposed to being described, it will have to be in terms of what is not alive; otherwise it must be regarded as an unexplained primitive.

Save for various provisional attempts to specify the defining characteristics of life for the sake of identifying and delimiting that which is to be explicated, biologists have tended to ignore the distinction between living and lifeless matter as being scientifically irrelevant. Life is something which biologists try to explain in terms of the nonliving. Finding no clear-cut boundary and no non-question-begging set of necessary and sufficient

conditions for a thing's being alive, many have been led
to deny the scientific usefulness of the concept altogether. None of the characteristics that are often cited
as being typical of life, such as growth and development,
self-replication, metabolic change with a constant flow
of energy, and the wholeness of an integrated system,
are totally lacking in the domain of the nonliving. Furthermore, the existence of bodies—viruses—which are
nothing but nucleic acid surrounded by a protein "coat"
but which in key respects behave as organisms raises
questions as to whether a molecule itself could be said
to be "alive." The concept of life does not signify any
simple property whose presence or absence can be clearly and unequivocally determined.

Physics and chemistry cannot specify the difference
between the living and the nonliving, not because the
difference is one that "transcends" physics and chemistry, but because these are not, nor were they ever intended to be, physical categories. Like the characterization of a continent in terms of political rather than
geological concepts, the classification of material things
in terms of living and nonliving represents the super-
position of a logically distinct set of categories whose
applications satisfy an entirely different set of needs.
The concept of life is obviously useful and important in
distinguishing living and dead chickens and in distinguishing hares from hunks of granite, but the biologist
who is committed to believing there to be a physical
basis to these distinctions wherever they occur need not,
and should not, assume that the categories themselves
must be specifiable in physicochemical terms. It is the
fundamentally misguided attempt to effect such a reduction, and the illegitimate extension of such terms as
'alive' beyond their accustomed domain to the domain

of molecules and suborganismic processes, that has led to much of the confusion over what the biologist means by "life."

There is no incompatibility between holding that organisms are living systems and that they are physical mechanisms with evolutionary histories. To say that organisms are machines and that they are alive is not to ascribe to them some occult or ghostly quality, but rather to indicate merely that they exhibit a loosely defined set of observable qualities that distinguish them from other machines which we would not describe as being alive. The expression 'living matter' is to be understood as signifying, not a special type of matter, or even ordinary matter specially arranged, but matter which happens to be situated in such a way as to bear a certain relation to a living organism. It is in this sense, and this sense alone, that it can be said that the organism confers life on matter, rather than the other way around.

5. The Extensions of Biological Thought

Biological systems are the entities about which, by which, and for which a science of biology has been constructed. The investigator himself is a paradigm of what biologists are concerned with. It is this peculiar relation of biological science to its subject matter that is responsible for many of biology's special philosophical problems. Biology is in the position of having to characterize man himself.

Biology is not, of course, the only science whose principles are applicable to the human organism, but it is the only one for which man represents an instance of its fundamental unit of classification. It is the biological individual that corresponds to the unit of personal identity. 'Man' is a biological predicate, and not a chemical or physical or psychological one. Psychology, it should be noted, can be construed either as a part of biology or as a discipline which abstracts from man and other organisms in order to characterize certain features which happen to be distinctive or most conspicuous in humans. Unlike biology proper, however, which treats man as one of a large number of classes of organisms to be described in terms of concepts not derived from the study of man at all, much of psychology seems to be at least implicitly based on the study of human behavior, with only derivative application to other organisms. If psychology were to succeed in inverting this anthropocentric order of knowing and to achieve a representation of the various characteristics of organisms ordinarily classified as "mental" as features alongside anatomical, physiological, and ecological properties—assuming this were possible—it would have essentially merged with biology.

Biology and the Idea of Progress

One of the ways in which the problem of man's position both within and without biology is manifest is in interpreting the course of evolution. Specifically, biologists have had to come to grips with the idea of progress in the context of phylogenetic change. Given that the authors of evolutionary theory, like all authors, are products of evolution, must their species be regarded as the culmination of the entire process? Evolution, it can be argued, is necessarily progressive, in the sense that any inheritable change that gives its bearer a selective advantage is progressive. Not all change is progress, but all change that has a survival value with respect to a selection procedure, whether natural or artificial, must be recognized as progress, since what is selected for is progressive by definition. Man, being the youngest mammalian species, must be the most highly evolved and hence the most progressive.

Such an argument falls far short of proving man to be the most progressive or most highly evolved species, however. In the first place, by making evolution *necessarily* progressive, it excludes nothing with respect to the nature of the process itself. It is fully compatible, for example, with the possibility that evolutionary change could be regressive in the sense that subsequent generations of organisms might turn out to be inferior in all respects other than the capacity to escape a certain type of environmental danger. One can easily imagine a situation in which all of the "best" (strongest, most intelligent, most perceptually sensitive, etc.) were systematically eliminated by some force which happened to spare only what by other criteria would be considered the least "advanced" specimens. Evolutionary change that followed this model could be called "progressive" only in a very narrow, technical, and uninformative sense.

Secondly, even if the claim that evolution exhibits progress were to be given empirical content, the argument would still fail to justify placing man at the apex of evolutionary progression. Man cannot be considered the most progressive, if only because he did not evolve either from insects or from seed plants. It can only be with respect to other species on its particular branch of the phylogenetic tree that a species can be considered highly evolved, if having a selective advantage is to be the basis for such comparison. It simply makes no sense in these terms to compare grossly divergent species with respect to evolutionary levels.

Sometimes the argument that evolution is progressive assumes a form that includes criteria of progress, so that the conclusion that man is the highest form of life can have a determinate empirical character. If biological change represents progress, it must be the case that later states are better than earlier ones; that is part of what progress means. The problem is one of finding suitable criteria or indices of evolutionary progress. Some of the criteria that have been suggested, such as brain size, ability to control the environment, and capacity for the acquisition and organization of knowledge, are blatantly question-begging. If progress is to be measured in terms of characteristics which are antecedently known to be either unique to man or to have reached their highest development in man, it should hardly come as a surprise to have it concluded that the crowning product of the evolutionary process is man. Arguments such as those which purport to establish the superiority of man over lower species by noting that only man has the mental capacity to frame the question (see, e.g., *56*, p. 256) are examples of this genre.

The character of the theory of evolution is such that any criterion of evolutionary progress that has been put forward is either arbitrary and hence question-begging, or else inadequate to accord man a unique position as the culmination of the entire evolutionary process. The criterion of relative dominance, for example, as Simpson points out (*130*, pp. 244-47), applies only to a restricted line of development and provides no basis for comparing the last and most progressive products of different lines of evolution. Increasing structural complexity, another possible criterion, is noted to be potentially misleading or irrelevant in cases where it has led to evolutionary "blind alleys" (*130*, p. 252). Short of adopting a stipulative definition, there is no way of defining biological efficiency or ability to adapt that would make mammals in general, and humans in particular, the most efficient or most highly adapted of the whole range of organisms. Progress as an evolutionary concept is meaningful only when relative to whatever preconceptions are held with respect to what is to count as the most advanced.

Considerations such as these have led some biologists to reject the notion of progress as having no place within a strictly scientific theory of evolution. Progress is an evaluative concept, and judgments of value are alien to the ideal of scientific objectivity. Biologists have often felt a certain uneasiness in speaking of progress, and it is perhaps significant that even Darwin, who clearly had thoroughly absorbed the nineteenth-century idea of progress, nevertheless wrote of the Descent—not the Ascent—of Man. No attempt to give the term 'progress' a purely biological or "objective" meaning is free from suspicion of contamination with evaluative connotation, and no concept that retains its evaluative features could be tolerated within a scientific theory.

Dispensing with the concept of progress entirely, however, may be an excessive price to pay for objectivity. Evolution is a historical process, and although one need not agree with Simpson that "it is impossible to think in terms of history without thinking of progress" (*130,* p. 239), it is important to recognize that it is essential to any theory of evolution that the changes be systematic rather than random or cyclical. Evolutionary change invariably involves trends, and these involve steps. The story of evolution is the tracing of a number of phylogenetic pathways; to trace such a pathway is to indicate a kind of progression. An evolutionary account is in a sense teleological, since it provides an interpretation of historical events as stages in the production of later states of affairs. A sequence of changes is described as progressive according as it occurs in the direction of some previously identified end. Progress presupposes a standard: not a moral standard, but a standard nonetheless. As long as the notion of progress is understood to signify development of a specific and antecedently conceived type, there is nothing subversively subjective or evaluative about using it.

Evolutionary biology not only admits of exposition in terms of the concept of progress, but demands it as well. Description of the processes of evolution requires a means of specifying sequences of changes, as opposed to individual changes. Simply to arrange such a sequence in an order presupposes some notion of progress. What gives the sequences of alterations that biologists investigate an evolutionary character is that they can be arranged in such a way as to constitute a progression. On the other hand, progress is also an empirical notion, in that it is always possible to discover that a given sequence of events is *not* progressive with respect to a particular criterion for progress. Evolutionary progress is evaluated on the basis of various measurable indicators of biological efficiency or adaptedness, and any series of changes can in principle be found to exhibit no progress by any of these standards. Progress is a category, or rather a class of categories, which is essential for providing an interpretation of phylogenetic changes as evolutionary processes, but it cannot guarantee that any such changes will in fact occur, or that phenomena will be found that will provide evidence for any evolutionary hypothesis.

Given the variety of ways both of defining and of achieving biological adaptedness and the multiplicity of progressive sequences of phylogenetic change, it follows that the theory of evolution by no means implies that man represents the highest level of evolutionary progress. It does not even imply that man is one of the highest products of evolution. Logically, it could be argued for any existing organism that it represents the culmination of the whole course of evolutionary history, since every existing form is adapted by virtue of certain features that it specifically possesses. The point is not that we can imagine a creature to have evolved far beyond human beings (the literature of science fiction provides some interesting possibilities of this sort), for this itself may presuppose the same type of prejudicial ranking of criteria that man's own self-elevation does; what is of key significance is rather that the respects in which one organism is better or more advanced than another are not naturally or supernaturally endowed with anything that sets them off from other respects that might yield an entirely different ranking. Criteria of progress are chosen, not given.

The "obviousness" of a scheme that places man at the top of the evolutionary progression or at the top of one of its principal branches might be compared to that of a geocentric conception of the universe. Unlike the latter case, however, in which the theory's replacement by one which saw man's planet dislodged from its former privileged position, the modern theory of evolution essentially leaves open the question of man's place in the total biological sphere. Investigations in the history of living forms does yield knowledge of sequences, but the theory itself does not provide a basis for attaching special importance to any of the various contemporary forms that occur as termini of the several lines of historical development. If any species can be said to represent progress, it can only be with respect to its extinct ancestors.*

*For further, more detailed discussion of the notion of evolutionary progress, see Goudge *47*, pp. 180-91, and Simpson *130*, chap. 15.

Biology, Mind, and the Behavior of Humans

More central to the general problem of biology's having to characterize biologists than the question of interpreting progress in evolution are the questions having to do with the implications of regarding a man as an organism at all. Obviously he *is* an organism, but is the biologist, at least qua biologist, committed to treating him as "merely" an organism? If man has any distinguishing characteristics, are these biological characteristics, and can they be analyzed and understood in terms of biological categories? Biology can reasonably be expected to handle the similarities between man and other organisms, but can it give an adequate account of the differences? It is here, where the boundary between biology and philosophy becomes obscured, that the metaphysical aspects of biological science come most sharply into focus.

If there exists a pervasive and fundamental presupposition of biologists concerned with the nature of man, it is that the human organism constitutes a proper object for the application of biological concepts. Seen from a biological perspective, man is *essentially* an organism, and it is as an organism that his origins, qualities, habits, and productions are to be understood. Accordingly, some biologists have held that the only sound approach to answering the question "What is man?" is to take the biological nature of man, both in his evolutionary history and in his present condition, as the only plausible point of departure. The emphasis is placed on those features of members of the human species which are continuous with characteristics of those of other species, and may be associated with a denial or a deemphasis of anything that would appear to reveal a basic difference between humans and other animals.

A striking, if somewhat extreme, example of this attitude is provided by George Gaylord Simpson, who, after acknowledging the profundity of the question "What is man?" and its central importance to any system of philosophy or theology, asserts that "all attempts to answer that question before 1859 [the year of publication of Darwin's *Origin of Species*] are worthless" (*129*, p. 80). Metaphysics, art, and literature can contribute nothing, he insists, "unless they accept, by specification or by implication, the nature of man as a biological organism," except as "fictional fancies or falsities." It is to be noted that it is not sufficient for Simpson that these other accounts be *compatible* with recognizing that man is the product of evolution. If such an account is to be of any value at all, it apparently must include some mention of man's biological nature, since, as a point of logic, no purely nonbiological account could have biological implications; that is why he regards all attempts made before 1859 to have been worthless.

Not all biologists subscribe to this position, of course, and it may actually be a rather small number that would be prepared to deny that the works of Plato, Aristotle, Sophocles, Dante, and Shakespeare tell us anything about the nature of man. The biologist does not *have* to accept only those accounts that stress man's shared animality. It does not contravene any biological principles to define the essence of man (or of any other species, for that matter) in terms of characteristics that may be species-specific. Nevertheless, there does exist a widespread tendency, one that stems from trying to think "biologically," to treat the various aspects of the study of man as part of biology, to be described in terms of biological functions and evolutionary origins. Human speech, tool-making, social behavior, and mental activity are commonly approached as adaptive features whose rudiments are to be found in other animals and whose peculiar intrinsic characters are secondary and derivative.

The procedure typically followed by those who approach the study of man "biologically" is to attempt to identify and describe the various distinctive features of his species in terms that have been used to characterize other types of organisms, and to trace their evolutionary origins by comparison with corresponding features of other organisms, both contemporary and extinct. In those cases in which there is much independent evidence of the occurrence of these features or their close analogs in other organisms, this approach has often been quite fruitful. It has been especially productive in the areas of anatomy, physiology, biochemistry, and embryology. Proceeding on the assumptions that all life processes are mediated by physicochemical mechanisms, that evolutionary trends are maintained only by natural selection, and that natural selection can work only on what is biologically useful to its possessors, biologists have achieved considerable success in coming to understand man's nature in terms of how his body is built, how it functions, and how it came to be the way it is.

With respect to some of man's other oft-cited distinctive characteristics, however, the contributions of the "biological" approach have not been so impressive. In particular, biologists' attempts to interpret man's mental characteristics have often been somewhat less than enlightening, at least to the extent that they have attempted to treat consciousness as a fundamental biological property that all organisms share. If consciousness, or awareness, is a property or aspect of any living thing, so it is argued, it must have evolved in the same way that all of the rest of its characteristics have. But there is nothing that resembles states of consciousness except other states of consciousness, so man's faculty of awareness must have evolved from more primitive levels of awareness to be found in creatures further down the phylogenetic tree. This is the position adopted by Sir Julian Huxley, who finds himself "driven to assume that . . . all living substance has mental, or we had better say mindlike, properties; but that these are, for the most part, far below the level of detection" (*66*, p. 96). Mental activity for Huxley—by which he means conscious awareness—has come into being and has reached the high degree it exhibits in humans solely as a result of the biological advantage it confers on its possessors, and can be traced back at least as far as the *Paramecium*, which he says "must be in some way aware of the difference between more acid and less acid water," and the *Euglena*, which "has a primitive awareness of light" (pp. 99, 100). Mind and matter, or the mental and the material, are conceived of as "two aspects of a single, underlying reality," and it is the mental aspect of life that "increases in importance during evolutionary time" (pp. 95, 93).

It is not difficult to see why Huxley was able to write an enthusiastic and laudatory introduction to Pierre Teilhard de Chardin's *The Phenomenon of Man*, which he affirms presents biological evolution as a process which "produces more varied, more intense and more highly organised mental activity or awareness" (*145*, p. 28). Teilhard, writing ostensibly as a biologist (in the Preface he insists that his book must be read "purely and simply as a scientific treatise"), is concerned with man *"solely* as a phenomenon," with stress on "the fundamental similarity of all organic beings" (pp. 31, 109). Like Huxley, he believes that there is a double aspect to the structure of the stuff of the universe, an inner or conscious aspect ("the *within* of things") and an external aspect ("the *without* of things"). Science, however, has heretofore been committed to the error of restricting the phenomenon of consciousness to the higher forms of life, so it is argued, and has failed to recognize consciousness as "a cosmic property of variable size subject to a global transformation" (p. 65). Noting that an outstanding characteristic of evolutionary development along that part of the phylogenetic tree that leads to man is the growth and complication of the nervous system, a process he calls "cerebralisation," Teilhard offers as an experimental definition of consciousness "the specific effect of organised complexity" (p. 329). "We may be sure," he affirms, "that every time a richer and better organised structure will correspond to the more developed consciousness" (p. 65). Consciousness, however, "transcends by far the ridiculously narrow limits within which our eyes can directly perceive it." "We are logically forced to assume," he concludes, "the existence in rudimentary form . . . of some sort of psyche in every corpuscle" (p. 329). Since *"nothing could ever burst forth as final across the different thresholds successively traversed by evolution . . . which has not already existed in an obscure and primordial way,"* one has no choice but to affirm that "by the very fact of the individualisation of our planet, a certain mass of elementary consciousness was originally emprisoned in the matter of the earth" (pp. 77-78; italics in original). Even molecules must be assumed to possess consciousness.

It is possible to discern two separate roots of Huxley's and Teilhard's conception of consciousness. One of these is the failure to distinguish between mental activity in the sense of feeling or conscious awareness and mental capacity as defined in terms of behavior. It is one thing to say that a creature (or an artifact) performs an act of discrimination or exhibits a reaction to a stimulus, and quite another to say that it is aware of what it is doing, regardless of the "level" of awareness that we are considering. Huxley conflates the two by referring to avoidance reactions and tropisms as forms of awareness, and even suggests that reflex actions result from various types of awareness (66, p. 100), ignoring the fact that such actions, at least in humans, are of precisely the sort that are recognized as automatic and that are assumed *not* to be invariably associated with consciousness. Teilhard falls into similar confusion when he cites the fly-trapping behavior of certain plants as proof that evolution within the vegetable kingdom, as well as within the animal kingdom, is "subservient to the rise of consciousness" (145, p. 169 n.). If the capacity for discriminatory response is logically sufficient for consciousness, then any robot, any automatic phonograph, and even any piece of litmus paper must be considered conscious.

The second root of the position that consciousness must exist, at least in rudimentary form, throughout the entire universe, consists in its reliance upon the argument that any property of an evolved entity must have been present in some form in all of its antecedents. The use of such an argument would imply not only that even the most primitive structures have a primordial degree of awareness, but also that they have a primordial sensitivity to color, sound, and smell. The ancestors of the hyena would have to have had a rudimentary laugh, and those of the lion, a rudimentary mane. Why an argument so patently fallacious seems to have gained widespread acceptance in the case of consciousness can perhaps be explained by the suggestion that in this instance it is supplemented by the implicit assumption that consciousness is an *essential* property of living matter, an assumption which, of course, begs the question. This suggestion may also explain why Huxley uses as an analogy the electrical properties of living substance (*66,* p. 96) and why Teilhard compares his imputation of imperceptible psychism with the physicist's attribution of mass to submicroscopic particles of matter (*145,* p. 330).

The difficulty with which all attempts to treat consciousness as a biological property are fraught is that consciousness is not a well-defined property the observation of which is equivalent to any indefinitely long series of observations either of behavior or of physiological states. There are no hard and fast criteria of states of consciousness in humans, much less for other creatures or for automata. No conceivable observation or deductive argument from empirical premises will constitute a *proof* of the existence of consciousness in anything other than oneself. We not only do not observe consciousness in lower forms of plant and animal life; we do not even observe it in other people. This is not to say that we do not or should not attribute it to other creatures, but only that it is not a diagnostic property. An object is called conscious if, and only if, it acts consciously; to act consciously is to behave in ways that have resemblance to certain biological paradigms and lack resemblance to certain nonbiological ones.*

*For greater elaboration of this point, see the author's *127.*

Because man is obviously a proper object for biological study, and because the property of having consciousness seems to be one of his most obvious characteristics, biologists have often felt a need to account for man's subjective awareness in evolutionary terms. Given the nature of consciousness, however, it is impossible in principle to show that it has evolved. For if consciousness can be detected, it must be more than an epiphenomenon, since it can be detected only by its effects: it would have to be causally efficacious. But whatever is recognizable as causally efficacious in an experimental sense is never a state of consciousness. What is conscious cannot be detected, and what can be detected cannot be conscious. Evolutionary biology can show the selective advantage, not of states of consciousness, but only of dispositions and capacities whose connection with consciousness falls outside of biology's purview.

To attempt to trace the evolution of consciousness is in some respects like trying to trace the evolution of honesty from other forms of life: it is to seek evidence in other species of that which perhaps can be significantly predicated only of humans. To speak of degrees of honesty presupposes the existence not simply of the possibility of certain rather complex behavior but of a set of social conditions and relationships that are unique to the human species. The same applies to consciousness: the terms 'conscious' and 'unconscious' have come into use not because we need them for the sake of designating certain behavioral capacities per se, but because they are required by the social institutions which human societies have created. Consciousness, to the extent that its foundations are essentially moral and legal, is a basically human concept, whose applications to nonhuman orders of being can only be derivative.

Whatever contributions a biological approach has made to understanding the "mental" characteristics of man have come forth from the study of behavior and its basis in physiology and biochemistry. Accepting the continuity of animal and human development which the theory of evolution implies, ethologists have begun to provide an intellectually useful base for understanding a number of man's voluntary and nonvoluntary activities. By comparing, for example, the male war-dance of the Siamese fighting fish with the ceremonial dances of Javanese and other Indonesian peoples, or the wolf's offering his neck to a stronger adversary, resulting in the avoidance of serious injury, with the Homeric soldier's throwing down his helmet and shield in an appeal for mercy (examples taken from *37*, pp. 155-60), the ethologist helps to render many aspects of human behavior intelligible by subsuming them under conceptual patterns which apply to a wide range of biological organisms, much in the same way that the anthropologist contributes to the understanding of the customs of his own culture by comparative studies of analogous customs in other cultures. The ethological study of man has provided not only an effective set of concepts for describing many of the things that men do but also a means of explaining these activities in terms of their functional and adaptive character.

On the other hand, the ethological approach to the human species has been afflicted with at least two general problems that have tended to undermine the value of studies of man which emphasize his similarities to other species. One of these problems is that this approach is systematically liable to the charge of failing to do justice to man's rather complicated and distinctive features, features which can only be understood in terms of their peculiarly human context. In particular, there are the "humanistic" aspects of man—his aesthetic, religious, and philosophical nature—which, if they are treated at all, are handled so loosely as to miss most of what is needed to understand them, like describing the activities of someone who makes his living speculating in stocks and bonds simply as an instance of food-gathering behavior. Desmond Morris's account of religion provides an example of this approach:

> As zoologists we must do our best to observe what actually happens rather than listen to what is supposed to be happening. If we do this, we are forced to the conclusion that, in a behavioral sense, religious activities consist of the coming together of large groups of people to perform repeated and prolonged submissive displays to appease a dominant individual. . . . The submissive responses to it may consist of closing the eyes, lowering the head, clasping the hands together in a begging gesture, kneeling, kissing the ground, or even extreme prostration, with the frequent accompaniment of wailing or chanting vocalizations. . . .
>
> At first sight, it is surprising that religion has been so successful, but its extreme potency is simply a measure of the strength of our fundamental biological tendency, inherited directly from our monkey and ape ancestors, to submit ourselves to an all-powerful, dominant member of the group. Because of this, religion has proved immensely valuable as a device for aiding social cohesion. . . . We simply have to "believe in something." Only a common belief will cement us together and keep us under control. [*96*, pp. 146-48]

It is not the accuracy of this portrayal that is to be

called into question, but its adequacy. If religion serves a biological need, it is a need whose distinctiveness seems to reside in its peculiarly human character and cannot be elucidated by an ethological (or pseudo-ethological) account. Metaphysical concern is not a biological property.

It is not only the humanistic side of man that may be slighted by the ethological approach, but also certain aspects of man as an object of scientific inquiry. One of these is the study of language and communication. W. H. Thorpe, for example, has sought to characterize human speech as "unique only in the way in which it combines and extends attributes which, in themselves, are not peculiar to man, but are also found in more than one group of animals" (*146,* p. 101). Human language he finds to be "propositional, syntactic and at the same time clearly expressive of intention," features which he also finds present, at least to some degree, elsewhere in the animal kingdom. The dance language of the bees is propositional, in the sense that it transmits precise information about the direction and distance of a food source. Evidence that animals can use conceptual symbols is provided by the recognition of numerals by birds. That animals can signal their intentions is indicated by the role of facial expressions in wolves and chimpanzees, as well as by such specific behavior as the intimidation of a subordinate male bison by a slight flick of a superior's head. Thorpe's conclusion is that "the distinction between man and the animals, on the ground that only the former possesses true language, seems far less satisfactory and logically defensible than it once did" (p. 102).

Explanations of human speech at the level of abstraction at which it resembles communication in animals, however, yields only a very superficial understanding of the nature of man's linguistic capacities. As Chomsky has pointed out, even walking can be seen to exhibit the properties Thorpe attributes to language, in the sense that he defines them: it is purposive in that it can express an intention, it is syntactic in the sense that it involves internal organization, and it is propositional in that it can signal one's interest in a certain goal by its speed or intensity (*23*, pp. 60-61). There is reason to believe that human language, unlike human gestural systems, is based on entirely different principles and operates by an entirely different mechanism from those involved in any form of animal communication. Whether or not it is true, as Chomsky believes, that "there is nothing useful to be said about behavior or thought at the level of abstraction at which animal and human communication fall together" (p. 62), the fact that human speech exhibits features which are common to all languages but which are not present in any other forms of animal communication makes it clear that there is a great deal more to explaining human language than comparative ethology is likely to discover.*

Another aspect of the human organism which the ethological approach has often oversimplified is man's social behavior in general, and his aggressive behavior in particular. Ethologists such as Konrad Lorenz, taking aggression to be a basic drive or instinct that man has inherited from his anthropoid ancestors, have sought to explain all of the various forms of human aggression (for example, wars) as manifestations of essentially biological tendencies. Faced with having "to realize how abjectly stupid and undesirable the historical mass behav-

*On the species-specificity of language and its basis in human biology, see Lenneberg *81*, esp. chap. 4.

ior of humanity actually is," Lorenz thus concludes that the reason "why reasonable beings . . . behave so unreasonably" is that human social behavior "is still subject to all the laws prevailing in all phylogenetically adapted instinctive behavior" (*83*, p. 229). Human behavior must be to a large extent biologically determined, he believes, since it is so often other than rational (meaning that it is not always consistent with the attainment of what a reflective person would presumably identify as its objectives). Man's social organization, Lorenz finds, is very similar in certain respects to that of rats, for example, whose aggressive behavior therefore provides at least one model for understanding human society. The implication he draws for man is that human aggression is something that cannot be avoided but can only be redirected in ways that will prove harmless to the species.

Conclusions of this sort have significant social implications, and whether or not they can be legitimately derived from ethological data is a question of considerable importance. Lorenz's position, which contains an implicit (and sometimes explicit) rejection of the possible importance of social and cultural factors in accounting for human aggression, has been the target of a number of criticisms from those who would offer an alternative interpretation of these data. One such alternative is the possibility that aggression is always a derivative phenomenon, a consequence of frustration, for example, brought on by environmental circumstances (see *33*). If this is the case, then the understanding of human aggression can be expected to demand an account in terms of the special sorts of stimuli and responses that are characteristic of humans, even though it also seems likely that ethologists can (and do) make contributions through comparative studies of various types of aggressive behavior.

Another suggestion, one that is much more threatening to the ethological approach to human aggression generally, is the view that what is most significant about man's social behavior is that it is not instinctive at all, in the sense of being stereotyped for the entire species, and that this is the fundamental difference between man and all other species (6). Man, after all, is the one species whose social behavior (except in infancy) does *not* depend on a uniform set of social signals. Barnett has pointed out that man's territoriality, unlike that of other animals, involves no pattern of signals or responses common to the whole species, by means of which unwanted visitors are turned away (5, p. 64). In man, the rules regarding property are culturally determined and vary widely in different human societies. Similarly, human rituals, unlike those involved in courtship among birds, for example, do not consist of a series of formal patterns of behavior which have evolved under the influence of natural selection (5, p. 107). Human beings share many instincts with animals, but what is most distinctive about humans may be their adaptability, their conspicuous lack of fixed patterns of behavior. It thus is entirely possible, and quite reasonable to suppose, that if a piece of behavior has to be learned, it cannot have evolved at all. If all that is biologically inherited is the capacity to be trained in various ways, then it can be wrong or highly misleading to construe the behavior patterns that are actually acquired as the results of phylogenetically determined instinctive mechanisms.

In addition to its vulnerability to the charge of oversimplifying man's special characteristics, the ethological

approach to the human species is plagued by its susceptibility to a vicious type of anthropomorphism. The
danger is that, in attempting to describe the behavior of
animals in terms whose primary role has always been to
describe the purposive behavior of humans, we may find
ourselves "reading" human characteristics into animals
and then inferring that we have much to learn about
ourselves from observations of these creatures. Carried
to a not too remote extreme, this process can have the
effect of providing illegitimate reinforcement of the
investigator's previous prejudices. A more subtle, and
probably more pervasive, form of anthropomorphism
within ethology concerns the selection of categories to
be used in characterizing animal behavior: there may be
a tendency to identify as supposedly unitary phenomena instances of behavior that have no common biological counterpart. The obviousness of one of these
categories may be either a function of the investigator's
peculiar social and cultural traditions, or it may be the
result merely of the mode of linguistic abstraction employed in picking out certain types of activities as fundamental. The identification of aggression and violence
as instinctive appears to be a case of this type of practice. Whatever the sociological bases for selecting these
as significant categories, inferring from the widespread
distribution of violent and aggressive behavior to the
existence of a fundamental biological drive may, from a
logical point of view, be no different from doing the
same for running or hurrying.

Another example of misplaced anthropomorphism
within ethology is the use of introspection as a means of
identifying the individual instincts. Man has just as
many instincts as he has qualitatively distinguishable
emotions, according to at least one defender of the in-

trospective method.* The model that is suggested is one in which the separate emotions are depicted as pure states of biological being, genetically determined and uniform throughout the species. This is a rather dubious conception of the individuation of human emotions, however. Neither introspective reports nor any other source provides any evidence whatsoever that there exist qualitatively distinct phenomena of consciousness corresponding to all of the subtly and extensively differentiated emotions specified in our language. We can explicate the difference between, say, grief and disappointment not in terms of a difference in introspectable feelings, but only in terms of the reasons or causes for being in such a state. What determines a particular emotion is the situation or environmental context in which it occurs. The reason a person cannot feel remorse for an action for which he knows he is not responsible, or feel jealous of a snowstorm, is not simply that these feelings and situations do not happen to be paired in that way, but that it is essential to the correct ascription and reporting of these or any other emotions that there obtain external conditions of a certain general sort. Emotion-words are not names of discrete feelings. And as for assuming that anything which is experienced as a strong drive must be instinctive, this implies a denial that our deeply felt attitudes are ever a function of learning. One is reminded of attempts to justify various human social practices, particularly with regard to the exploitation of others, in terms of "human nature." The problem with a classification of instincts based on introspection is not so much that the anthropomorphism inherent in this method would itself compromise the ethological ap-

*Lorenz. See Fletcher 37, p. 155.

proach, as it is that it might tend to generate an ethology that could not be supported biologically, given the fact that our experienced drives often have social and cultural determinants. Anthropomorphism is not bad ethology, but it can produce bad ethology, which in turn may produce bad anthropology.

Up to this point, the question of the value of ethological studies as a means of helping us understand human social behavior has been treated in basically scientific terms. The pitfalls to which I have tried to call attention are essentially impediments to producing good science, as opposed, not to no science at all, but to poor science. Quite independent of the question of whether Lorenz's or anyone else's conclusions are warranted by the evidence adduced is the question of whether studies of animal behavior are capable of generating any objective conclusions at all concerning man's social behavior. Can a science of ethology ever be expected to yield insight into the nature of human society, or are all such efforts to ground social science in biological fact doomed in principle to failure?

I believe that one has no choice but to be deeply skeptical of the possibility that ethology can provide an effective basis for the objective study of man's social nature. In the first place, it seems unlikely, if not impossible, that even the controversy over whether man is inherently aggressive can ever be settled. We know that not everyone is overtly aggressive. What would ultimately decide whether aggression is a basic drive that is often masked or repressed, or whether it is something that is produced when other drives are thwarted or when certain types of stimuli are present? Any kind of behavior, whether instinctive or not, can be prevented from occurring, given sufficiently drastic means. We know that au-

thoritarian and permissive upbringing of children produce different results, but can we decide which results
are "natural"? What sorts of behavior is it "natural" to
suppress? It is easy to see that debate over these questions must ultimately collapse into a dispute over values
and over man's philosophical essence. Though it will be
granted that the amenability of a given set of data to
accounts by a multiplicity of theories by no means rules
out the possibility of the emergence of a single theory
that outstrips all of its competitors, the situation with
respect to human aggression seems to be one which no
amount of data could ever resolve.

A second reason for doubting the soundness of ethologists' claims to be making contributions to the understanding of human social behavior has to do with the
nature of the explanations they might be able to offer.
On the one hand, a purported explanation of a segment
of human behavior may take the form of simply showing it to be similar to something that animals do, the
alleged content of such an explanation being that the
behavior has been demonstrated to be instinctive and
hence biologically determined. What is most unsatisfactory about an account as superficial as this one is
that it is defenseless with respect to the charge of failing
to provide adequate grounds for saying that the human
and animal behavior really are sufficiently similar to
warrant ascribing similar causes to them. If, on the other
hand, a common biochemical substratum mediating the
respective activities is also found, what we would then
have would be a physiological or biochemical explanation, not an ethological one. Comparative studies would
have been useful only as a means of investigation, much
in the same way that experiments involving the use of
animal subjects are useful in carrying out medical re-

search. Studies of animal behavior would have the role merely of providing data for the physiological and biochemical sciences to explain.

We cannot, of course, rule out a priori the possibility that an increased understanding of the biological bases of behavior might turn out to be extremely useful in explaining human social behavior, even though it might seem extremely unlikely. If, for example, imputedly analogous segments of human and animal behavior X_m and X_a are both found uniquely associated with a common molecular substance S, there is good reason to suppose that X_m is in some sense innately determined if X_a is. Since human behavior, as we have noted, does not consist of rigidly determined, fixed patterns and in fact exhibits considerable variation throughout the species, even when it is supposed to be instinctive, findings of this sort can have explanatory force only where these variations are deemed unimportant. The problem is that those portions of human behavior for which this is true—eating, drinking, and sexual behavior, for example —tend to fall outside the area of interest for social science. Many, if not all, of the kinds of phenomena which we would like to have explained by a social theory— racial prejudice, homosexuality, wars, personal and group rebellion, social control—turn out to be cases in which it is the *differences* among the various manifestations of a given drive that interest us. To the extent that social categories consist of items such as individual personality traits and aspects of behavior, items which are identified only in terms of a network of human institutions that they presuppose, they simply do not apply, except derivatively, to other species, and hence cannot be derived from any science of ethology.

Social behavior among humans can almost never be

simply biological. The discovery of the XYY chromosomal irregularity and its connection with crimes of violence is certainly illuminating, but not only will this not
explain criminal behavior in general, it will not even
explain the cases in which it is relevant, in the sense of
providing an account that is sufficient to show why the
deviant behavior caused by the anomaly produces the
particular manifestations that a specific case may present. Murder, rape, and armed robbery are complex social phenomena, and the causation cannot be attributed
to a single circumstance, whether biological or otherwise. The most that any discovery in biobehavioral science could show is that certain people, or all people,
have an innate predisposition to exhibit certain types of
social behavior (such as fighting of various sorts) or to
develop certain attitudes (such as racial prejudice), given
the appropriate sets of institutions and social stimuli.
Biology may be able to discover one or more of the
necessary conditions for behavior of a given sort, but
unless we deny that behavior is ever influenced by sociocultural factors, it can never be expected to provide
sufficient conditions for all of human social behavior.
Biological science can no more do the job of social science than geology can do the job of astronomy.

Biology and Ethics

A topic in human affairs that has been a perennial
source of interest to a number of distinguished biologists has been that of ethics. It is also a topic that displays both of the general problems that have haunted
attempts to view man's nature from a biological perspective. Inquiry into the biological basis for ethics has been
anthropomorphic not merely in the sense that it has

involved designating something important and essentially unique to man as a character to be treated biologically, but also in having produced efforts to find instances of moral sense in other animals and attempts to derive ethical norms from knowledge of animal behavior and evolution. Darwin, for example, in *The Descent of Man*, claimed to have discerned "something like a conscience" in dogs, and argued that the moral sense develops normally upon development of the intellect. The mark of human savages was supposed to be their low morality. Civilized man, by virtue of the selective advantage conferred on him by his higher morality, could be expected to exterminate and replace the savage races (*28*, chaps. 4-5).

A conception of evolution that yields ethical views such as those of Darwin and the social Darwinists is not merely anthropomorphic, but is conspicuously parochial as well. There is as much basis in the biological theory of evolution for justifying an equal distribution of land and resources throughout the human species as there is for justifying ruthless competition, laissez-faire capitalism, and political imperialism.* One set of conclusions follows from interpreting Darwinian theory as prescribing protection of all members of the species and encouragement of diversity and variation; the other follows from taking the struggle for survival as a model for justifying selfishness. Since Hume, however, it has been recognized to be a fallacy to suppose that a conclusion about what ought to be can be deduced from premisses about what is, was, or will be. The theory of

*For illustrations of the ease with which protagonists of various moral and political points of view have been able ostensibly to derive mutually incompatible propositions from Darwin's theory, see Flew *39*, pp. 5-6, 36-37.

evolution does not tell us what is good or bad, in any ethical sense; the evolutionary process itself is non-ethical.*

Even if it were to be assumed that evolution has a direction, that human welfare and survival of the human species are its goals, and that human evolution is mediated by ethical beliefs, however, it would still not be possible to derive a system of ethics from evolutionary theory. There are too many additional variables for a single coherent set of ethical principles to be so determined. There is ordinarily no way of knowing, for example, when it will be most advantageous to tolerate individual differences at the expense of the group, and when the preservation of the group is to be given precedence. It makes no sense to speak of favoring that which is adaptive biologically without specifying what sorts of conditions it is to be adapted *to,* and this can rarely be anticipated. Furthermore, as T. H. Huxley pointed out long ago (*68*), ethical beliefs may not even be compatible with evolutionary principles. Caring for the aged and infirm members of a society is an example of ethical behavior that may not actually benefit the species at all. In any case, human ethical systems, whether adaptive or nonadaptive biologically, always involve supplementary ethical principles whose desirability cannot be assessed in terms of the furtherance of evolutionary goals.

The providing of an ostensibly biological account of man's ethical nature has to a large extent been made possible by succumbing to the other tendency to which those who seek to apply biological thinking to human

*That is why someone like Julian Huxley needs to base his evolutionary ethics not just on evolution but on his conception of the nature of man, which in turn depends on an anthropomorphic conception of evolutionary progress; see Huxley *67;* also Simpson *129,* p. 142.

problems are prone, namely, that of emphasizing similarities between humans and other animals at the expense of ignoring important dissimilarities. Biologists appear to have been led into considering man's ethical characteristics as products of evolution as a result of treating the development of ethical systems as a kind of evolution, akin to if not actually a part of biological evolution. Recognizing that the early versions of evolutionary ethics, such as that of Herbert Spencer, "have been vitiated through their attempting to apply conclusions derived solely from the biological level of evolution to subjects like ethics which only come into existence on the social level" (*67*, p. 79), biologists have sometimes moved to the view that man's ethical character is a product of a special type inheritance that only humans share: cultural or social inheritance. Man has a method for transmitting potentialities to later generations that does not involve either genes or sex cells: teaching. Cultural evolution is the process that is responsible for the changes in man's conceptual knowledge, beliefs, aesthetic creations, and other artifacts, as well as his ethical ideas of right and wrong.

In treating cultural evolution as essentially continuous with biological evolution, however, biologists have sometimes obscured the distinctions between phenomena which fall under the rubric of biology and those which do not, and biological terms come to verge on the metaphorical. Julian Huxley, for example, speaks of "a new type of organization [that] has come into being—that of self-reproducing society" (*67*, p. 38). Societies become the bearers of characters, analogous to organisms, and moral codes are described as essential functional parts, not of individuals but of social organizations. C. H. Waddington (*153*), on the other hand,

226   *The Matter of Life*

though sensitive to the important disanalogies between
human societies and animal organisms, and sharply critical of attempts to treat the former as a species of the
latter, nevertheless also falls subject to a facile con-
flation of biological and cultural inheritance. Thus by
comparing the "socio-genetic" transmission mechanism
with various modes of "para-genetic" transmission in
animals, such as the passing of substances into the offspring of mammals either from the maternal blood
through a placenta or through the mother's milk, he
comes to interpret the transmission of culture as "an
enormous expansion and multiplication of modes of
para-genetic transmission" (p. 113). The most important
features of sociogenetic transmission that Waddington
cites to distinguish it from paragenetic transmission are
that it can occur throughout the greater part of the
individual's life, and that it handles groups of units
linked together by their functional interactions, as in
the case of a complex industrial technique or set of
religious doctrines. What he neglects to mention is that
paragenetic transmission occurs only in cases in which
there is a *biological* relationship between parties involved, and that the determinants of whether anything
actually is transmitted are always biological, whereas
culture can be transmitted to any number of contemporaneous individuals or to future generations, regardless of biological descent or relationships. The factors
which determine whether such transmission occurs are
ordinarily not biological, or at least need not be defined
in biological terms. Furthermore, although paragenetic
and sociogenetic transmission share the feature that
what is transmitted may or may not be passed on to
subsequent generations, it is perhaps significant that the
former is not ordinarily a form of inheritance at all

Cultural inheritance is interpretable as a form of inheritance only in an extended, metaphorical sense. Cultural evolution must be regarded as a type of evolution in which nothing is inherited.

Ethics, even descriptive meta-ethics, cannot be a part of biology, because cultural evolution is not simply an advanced form of biological evolution. This is not to deny that evolutionary notions provide useful analogies for understanding cultural phenomena, but only to deny that these phenomena constitute actual instances of evolutionary processes in anything like a biological sense. The evolution of cultural institutions and characteristics is no more a biological process than is the "evolution" of a house which a man builds and bequeaths to his son, who modifies it and bequeaths it to his son, who subsequently modifies it, and so on.

Because ethics is species-specific, it is anthropocentric and anthropomorphic. Ethical principles cannot be derived from biology, not merely because this would involve the naturalistic fallacy of trying to derive "ought" from "is," but because propositions containing terms that apply only to man cannot be deduced from propositions in which these terms are absent. What one can do as a biologist is to ask how evolution can have produced an organism that is capable of cultural evolution and the development of ethical systems. Providing an answer to this question, however, does not give us a system of ethics.

Biology and Its Applications

The question of biology's relation to ethics can be viewed as part of a larger, more general question, that of the relation of biology to its applications. Man is a biological organism that is capable of using biological knowledge as a means of influencing the future of his species, not only by altering his environment but also directly. Medicine, agriculture, and eugenics all represent ways in which man can manipulate his own and other species for the sake of achieving various goals. The pursuit of biological knowledge, as an end in itself, may not be much illuminated by treating it in biological terms; to classify it as a type of exploratory behavior does not go very far in explaining this peculiarly human activity. The *using* of biological knowledge, however, is a topic that offers somewhat greater possibilities for a biological interpretation.

Thinking biologically, one would expect those who apply biological knowledge to be concerned with the preservation not of life in general but rather with that of their own species, often at the expense of other species. Medicine and the development of antibiotics, for example, can easily be seen as means of promoting a selective advantage for the human species. The same would seem to hold for applications to agriculture, the development of insecticides, and human ecology. In other cases, such as the care and training of the mentally and physically handicapped and the attention paid to other species whose continued survival has no known positive effect on that of the human species, the contributions of biological knowledge toward maintaining the human species are not so apparent. The justification of such use of human resources typically involves only the appeal to human values, especially since any biological explanation—for example, as vestiges or rudimentary traces of habits having some adaptive significance among pre-existing forms—would be extremely weak. Still other ways of exploiting biological knowledge, such as the development of methods for carrying out biological warfare, seem to offer no advantage to the species, but only to a subgroup within that species, if to anyone at all. It is always possible to put forward an explanation of such behavior (by interpreting biological weapons as the means by which the "fittest" members of the species gain ascendancy in the struggle for survival, for example)—that is what makes the application of biological notions to human affairs often seem so empty—but in this case, as in many others, explanations in terms of nonbiological ends are generally more informative. If everything the biologist does is construed as beneficial to the species, then to say of something he does that it is, is to say nothing at all.

If it is assumed, nevertheless, that the proliferation of biological knowledge and its applications corresponds to increasing adaptedness of the human species, we are still faced with the problem of anticipating the species' future needs and determining the direction of scientific research accordingly. Controversies inevitably arise concerning not only which groups within human society should be benefited, but also what will be most useful to the species as a whole. In the first place, there is the problem of deciding which human variants should be fostered in establishing, for example, priorities in types of medical research—whether it is more beneficial to concentrate on organ transplants or on congenital blood diseases. Secondly, there is the problem of determining what sorts of immunities, skills, and structural characteristics will turn out to be advantageous under future conditions of terrestrial life. It is not ordinarily possible to predict what kinds of events or catastrophes may befall the earth, whether cosmic or man-caused. Even if we always do the best we can, we must recognize the possibility that what we do may be wrong, in the sense of being maladaptive, and there is no guarantee, based on biology or on anything else, that the views that prevail in these controversies will turn out to ensure biological success to the human species. There is no reason to believe of any species that it is destined for immortality.

These matters assume an especially acute form when it comes to the possibilities of determining human genotypes and phenotypes through genetic and biochemical manipulation. For in addition to the problems of our inability always to anticipate adequately the demands future environments will impose on the earth's possible inhabitants, there is the more fundamental human question of the implications of being able to predict and determine the genetic constitution and behavioral capacities of members of our own species. Supplemented by the knowledge of how to determine behavior by control of the environment and by the introduction of chemical substances into the body, a program could in principle be devised and carried out that would guarantee an order to human society whose nature could be precisely known and controlled. Given an intraspecific (i.e. nonbiological) understanding of the nature of man, some would argue that the realization of such a possibility would in itself alter the human species. It is of the essence of man, so it might be maintained, that should be impossible in practice, if not in principle, fully to predict and control his behavior and attitudes. Whether or not such a definition of man's nature is actually adopted, it does seem clear that the completion of the program envisioned would mark a significant alteration in the nature of the human species.

The approach to man which concentrates on those of

his features which are species-specific might broadly be called "humanism," even though it can be "scientific" as well. What it is not is "biological," in the sense of requiring the use of concepts that are necessarily applicable to other organisms as well. To the extent that man is regarded as sui generis as an object of description, there is no single "biological" point of view, but rather a number of quite different points of view, many of which are entirely compatible with biology. Disciplines which are included among these, some of which may have nothing at all to do with biology, are philosophy, religion, history, sociology, linguistics, much of psychology, and studies of man's artistic creations. They are also what perforce are left out of a strictly biological account of man.

Man is, among other things, an instance of a biological species, but even this fact is susceptible to a variety of approaches, since a species may occupy any of a large number of possible positions within a biological perspective. It is no more of a biological principle that the survival of any particular species is desirable than it is that its extinction is desirable. It does not even follow from biology that we as human organisms should even *want* ourselves or our species to survive. Whether or not our wants are determined by natural selection, it follows neither that we shall achieve what we want (even assuming there were universal agreement as to what the members of the species *do* want), nor that we shall survive even if we do achieve what we want. Biology offers no assurance that man's cultural products will ultimately prove adaptive, or that the human species does not represent the end of an evolutionary blind alley.

Conclusion

A science of biology is possible because man is able to discover order among and within living things. Organisms are classifiable and hence describable, and their internal workings have been found to reveal sufficient coherence and uniformity to allow systematic accounts to be constructed. There are enough characteristics and processes that the things we call alive have in common, and enough discernible order on the level of their interior parts, to make detailed description, analysis, and comprehensive schematization possible for this distinctive mode of being.

Biological description, like any other form of description, involves abstracting certain features of the object being described. To describe is not to *reproduce* a state of affairs, but merely to represent it. Accordingly, an adequate description is one that leaves out that, and only that, which is inessential to understanding the phenomenon under consideration. This, of course, depends on the type or level of understanding that is sought. Physics, for example, offers descriptions which abstract from all features, applications, and levels of organization other than those which are expressed in terms of mass, charge, energy, electric field strength, and so on. Reductionist or mechanistic accounts of biological organisms have sometimes been claimed to be capable in principle of explaining life processes in purely physicochemical terms and of providing definitions of all biological expressions in terms of physicochemical ones. Biology has a useful level of abstraction, which is not the same as that of chemistry or physics, regardless of whether biological expressions are eliminable or not. The issue of reduction is not so much one of logical derivability as one of the pragmatic relevance of attempting to dispense with certain levels of abstraction.

We have seen a number of ways in which the behavior of biological material has strained the mechanistic ideal of providing accounts of biological processes in strictly causal terms. Indeed, the logical development of biological science can be viewed as an expansion and extension of modes of description to meet the special demands which biological subject matter imposes. Biology's use of historical, functional, and teleological types of explanation represents a response to a challenge posed by organized systems in general and biological systems in particular: the challenge to provide descriptions which do not abstract from those characteristics the grasping of which is essential for understanding the system but which would not be included in a description on the level of chemistry or physics. Biology resorts to accounts of this sort, not because a causal or mechanistic account of a life process cannot always be produced, but because they are demanded by the nature of the systems studied and the type of understanding sought.

Are there limits to the type and level of understanding of living things that biology can provide through its conceptual apparatus? Is there anything one would want to know about a living system that biology cannot contribute? The answer to both questions is obviously yes, as we have seen in the cases of consciousness, ethics, and the humanistic disciplines. Biological science grades into social science, to be sure, but the latter also has produced its own concepts as means of dealing with those levels of organization with which it deals, levels from which the biologist ordinarily abstracts. The question of what social science itself, as it is variously conceived, leaves out is still another matter that needs to be considered, but not one that can be addressed here.

All science, as has been remarked, is abstraction. Biological science, however, by virtue of being directed toward things that are not only systems but also living, and because of its pivotal position between the physical sciences and the human sciences, has always been beset with a peculiar problem involving a deep controversy between two competing modes of abstraction: the dispute between mechanism (and its descendants) and vitalism (and its descendants). The mechanist (or reductionist or molecular biologist) is one who abstracts that which can be rendered in physicochemical terms from whatever organization an object possesses above the chemical level, whereas the antimechanist (or antireductionist or organismic biologist) is one who abstracts the holistic, organizational, purposive, and other characteristically biological features from that which the mechanist views as an arrangement of molecules. The former is accused of oversimplifying and failing to see the forest for the trees, the latter of impeding the advancement of biological inquiry by rejecting analysis as a route to understanding. The issue is one over the primacy of biological or physicochemical conceptualizations, whether the question of what it means to say that an organism is alive can best be answered by abstracting from the fact that it is alive.

The classic controversy between mechanism and vitalism, it has been argued, as well as the contemporary dispute between the molecular biologists and the "organicists," or the reductionists and the antireductionists, is an instance of a metatheoretical disagreement rather than a substantive one, because it is not resolvable by appeal to scientific evidence (58). It is viewed as an issue that is based, not on different findings and expectations as to experimental results, but on differences in fundamental philosophical commitments. The signs of such a dispute, it is suggested, include a large polemic component, methodological disagreement, disagreement over what is to be counted as evidence, and the use of a defense in terms of the heuristic value of a position. Whenever these features are present, so it is maintained, what we are dealing with is not a theoretical issue but a metatheoretical one, a manifestation of the sort of commitment that determines life style, political attitude, and religious point of view.

It is plain that the disputes between mechanist and vitalist or the reductionist and the antireductionist— between the "nothing but" people and the "over-and-above" people—do reveal fundamental differences in outlook, and that they exhibit all of the characteristics typical of a defense of one's commitments. In the case of the dispute between the reductionists and the antireductionists, the quarrel seems to be over which is "more basic" and which is "more important," the biological level of organization or the chemical one. In some instances it is the value of a scientist's work that is at stake; in others it may be the soundness of a metaphysical view concerning the nature of man or of life on earth. Whether the issue comes down to one of deciding between opposing nonempirical claims (such as assertions concerning the existence of a substantive vital principle) or of determining what aspects of a living system should be emphasized in seeking to understand biological phenomena, the disagreement is one that depends on differences of a personal, psychological, or sociocultural nature.

Nevertheless, it may be quite misleading to designate such controversies as metatheoretical, as if to suggest either that they are not really scientific or that a truly scientific controversy does not include such factors. *All* scientific disputes involve an element of commitment, and the waging of controversies very frequently involves polemic, challenges as to the relevance of and weight assigned to various pieces of evidence, and appeal to heuristics. If one looks not merely at the classic controversies in the history of science—controversies which have made significant intellectual history *because* of the philosophical commitments involved—but to any of a number of smaller disputes within science—the controversy between the crossing-over and reduplication theories of genetic recombination, discussed in chapter 3, is only one example—one sees that these disagreements have a "metatheoretical" character to them as well. It would be naïve to suppose that theoretical beliefs on any level are solely and uniquely determined by scientific evidence, or that factual criteria are the only criteria by virtue of which scientific theories are accepted, and it would be absurd to deny that those classic cases in the history of modern science in which fundamental conceptions of the nature of the physical world were challenged were in some sense not really scientific disputes. Clash between opposing commitments is the hallmark, not the antithesis, of a genuine scientific controversy.

Polanyi's stress (*107,* esp. chap. 8) on the importance of commitment in understanding the creation, adoption, defense, and change of scientific theories is well taken. The scientist's beliefs are not strictly and mechanically generated from empirical data, nor do they change in any such manner. As is the case with respect to political beliefs and religious views, a scientist's expectations and interpretations of new experiences are a function of his primary commitments, as are his responses to additional information and assertions. Science is methodical, of course, but no application of formal rules of proceeding and no appeal to objective evidence are sufficient to account either for the scientist's commitments or for the behavior that results from them.

Commitments, on the other hand, are not absolute, and while it is clear that the growth and development of theory presupposes the making of choices, it does not follow that these commitments are not susceptible to change. One may distinguish as levels of commitment predisposition, bias, and conviction, and none of these is necessarily immutable. There are indeed such things as conversions in all areas of commitment. Science is a domain in which there is a mechanism for changing one's commitments, albeit not always a very effective one: the appeal to evidence. Scientific controversies vary in their susceptibility to adjudication by this or any other means, but that feature need make them no less scientific, so long as the respective positions are based on empirical evidence.

The mechanist-vitalist debate, along with its contemporary descendants, is an instance of a scientific dispute in which the role of scientific evidence has been relatively small (although the historical changes in the respective positions have largely come about as a result of experiments). This is not to say that the dispute is or is not a philosophical or metatheoretical one, however. The point is that there is no such thing as a strictly scientific level of disagreement, as distinct from a philosophical one. There is something like a continuum, ranging from the predominantly empirical to the purely conceptual or theoretical: a controversy is more or less "philosophical" according to the depth of the principles that are being challenged. To the extent that science and philosophy both are concerned with the question of what we are to make of facts, there is no clear distinction between them. The dispute between the two approaches to biology is both scientific and philosophical because it depends on a fundamental difference in ways of conceiving the world of living things. But its significance with respect to the nature of biological science resides not in the depth of philosophical commitment involved, or in the height of abstraction from empirical observations at which it exists, but in its centrality, the fact that it comes forth at the very starting point of the science. The two competing attitudes represent radically different answers to the first question of biology: What is a living thing?

Like physics, biology is a part of natural philosophy: it gives us a means of understanding a portion of what is naturally given. What biology seeks to comprehend are objects that are alive. These objects exhibit a character that places them at a point between the natural and the artificial. Biology is concerned with that which displays an order that gives the appearance of having been designed, and yet at the same time performs as something man could never have designed. The organism is at once an inscrutable manlike object and a soluble puzzle of intelligible science. It is that tension that is responsible for the special character of the science of living things.

Bibliographical References

1. Achinstein, Peter. "Models, Analogies, and Theories," *Philosophy of Science* 31 (1964): 328-50.
2. Achinstein, Peter. "Theoretical Models," *British Journal for the Philosophy of Science* 16 (1965): 102-20.
3. Ackermann, Robert. "Mechanism, Methodology, and Biological Theory," *Synthese* 20 (1969): 219-29.
4. Avery, O. T.; MacLeod, C. M.; and McCarty, M. "Studies on the Chemical Nature of the Substance Inducing Transformation of Pneumococcal Types," *Journal of Experimental Medicine* 79 (1944): 137-57.
5. Barnett, S. A. *Instinct and Intelligence.* Englewood Cliffs, N.J.: Prentice-Hall, 1967.
6. Barnett, S. A. Review of *On Aggression* by Konrad Lorenz. *Scientific American* 216, no. 2 (February 1967): 135-38.
7. Bateson, William. *Mendel's Principles of Heredity.* Cambridge: Cambridge University Press, 1909.
8. Bateson, William, and Punnett, R. C. "On Gametic Series Involving Reduplication of Certain Terms," *Journal of Genetics* 1 (1911): 293-302.
9. Beckner, Morton. *The Biological Way of Thought.* Berkeley and Los Angeles: University of California Press, 1968.
10. Benzer, Seymour. "The Elementary Units of Heredity." In W. D. McElroy and Bentley Glass, eds. *The Chemical Basis*

of Heredity (Baltimore: Johns Hopkins University Press, 1957), pp. 70-93.

11. Bernal, J. D. *The Origin of Life*. Cleveland: World Publishing Co., 1967.

12. Bertalanffy, Ludwig von. *Modern Theories of Development*. Translated by J. H. Woodger. New York: Harper & Row, 1962.

13. Bertalanffy, Ludwig von. *Problems of Life*. New York: John Wiley & Sons, 1952.

14. Blum, H. F. *Time's Arrow and Evolution*. Princeton: Princeton University Press, 1951.

15. Bohr, Niels. "Light and Life," *Nature* 131 (1933): 421-23, 457-59.

16. Braithwaite, R. B. *Scientific Explanation*. Cambridge: Cambridge University Press, 1953.

17. Bunge, Mario. "The Weight of Simplicity in the Construction and Assaying of Scientific Theories," *Philosophy of Science* 28 (1961): 120-49.

18. Calvin, Melvin. "Origin of Life on Earth and Elsewhere." In *The Logic of Personal Knowledge* (London: Routledge & Kegan Paul, 1961), pp. 207-30.

19. Canfield, John. "Teleological Explanation in Biology," *British Journal for the Philosophy of Science* 14 (1964): 285-95.

20. Carlson, E. A. *The Gene: A Critical History*. Philadelphia: W. B. Saunders Co., 1966.

21. Causey, Robert L. "Polanyi on Structure and Reduction," *Synthese* 20 (1969): 230-37.

22. Chargaff, Erwin. "Chemical Specificity of Nucleic Acids and Mechanism of Their Enzymatic Degradation," *Experientia* 6 (1950): 201-09.

23. Chomsky, Noam. *Language and Mind*. New York: Harcourt, Brace & World, 1968.

24. Commoner, Barry. "In Defense of Biology," *Science* 133 (1961): 1745-48.

25. Correns, Carl. "Scheinbare Ausnahmen von der Mendelschen Spaltungsregel für Bastarde," *Berichte der Deutchen Botanischen Gesellschaft* 20 (1902): 159-72.

26. Correns, Carl. "Ueber den Modus und den Zeitpunkt der Spaltung der Anlagen beiden Bastarden vom Erbsen-Typus," *Botanische Zeitung* 60 (1902): 65-82.

27. Crick, F. H. C. "The Origin of the Genetic Code," *Journal of Molecular Biology* 38 (1968): 367-79.

28. Darwin, Charles. *The Descent of Man*. 2 vols. New York: J. A. Hill & Co., 1904.

29. Darwin, Charles. *On the Origin of Species.* A Facsimile of the First Edition. New York: Atheneum, 1967.
30. Darwin, Charles. *The Variation of Animals and Plants under Domestication.* 2 vols. New York: D. Appleton & Co. 1897.
31. Davidson, Donald. "Mental Events." In Lawrence Foster and J. W. Swanson, eds. *Experience and Theory* (Amherst, Mass.: University of Massachusetts Press, 1970), pp. 79-101.
32. Delbrück, Max. "A Physicist Looks at Biology," *Transactions of the Connecticut Academy of Arts and Sciences* 38 (1949): 175-90.
33. Dollard, John; Doob, Leonard W.; Miller, Neal E.; Mowrer, O. H.; and Sears, Robert R. *Frustration and Aggression.* New Haven: Yale University Press, 1939.
34. Duhem, Pierre. *The Aim and Structure of Physical Theory.* Translated by Philip P. Wiener. Princeton: Princeton University Press, 1954.
35. Dunn, L. C. *A Short History of Genetics.* New York: McGraw-Hill, 1965.
36. Feyerabend, P. K. "Explanation, Reduction, and Empiricism," in H. Feigl and G. Maxwell, eds. *Minnesota Studies in the Philosophy of Science,* vol. 3 (Minneapolis: University of Minnesota Press, 1962), pp. 28-97.
37. Fletcher, Ronald. *Instinct in Man.* New York: Schocken Books, 1966.
38. Flew, A. G. N. " 'The Concept of Evolution': A Comment," *Philosophy* 41 (1966): 70-75.
39. Flew, A. G. N. *Evolutionary Ethics.* New York: St. Martin's Press, 1967.
40. Flew, A. G. N. "The Structure of Darwinism," in *New Biology,* no. 28 (Baltimore: Penguin Books, 1959), pp. 24-44.
41. Fox, S. W., ed. *The Origins of Prebiological Systems.* New York: Academic Press, 1965.
42. Frankfurt, Harry G., and Poole, Brian. "Functional Analyses in Biology," *British Journal for the Philosophy of Science* 17 (1966): 69-72.
43. Glass, Bentley. "The Establishment of Modern Genetical Theory as an Example of the Interaction of Different Models, Techniques, and Inferences." In A. C. Crombie, ed. *Scientific Change* (New York: Basic Books, 1963), pp. 521-41.
44. Glass, Bentley. "Maupertuis and the Beginning of Genetics," *Quarterly Review of Biology* 22 (1947): 196-210.
45. Goldschmidt, R. B. "Different Philosophies of Genetics," *Science* 119 (1954): 703-10.
46. Goldschmidt, R. B. *Physiological Genetics.* New York: McGraw-Hill, 1938.

47. Goudge, T. A. *The Ascent of Life.* Toronto: University of Toronto Press, 1961.

48. Grant, Verne. "The Development of a Theory of Heredity," *American Scientist* 44 (1956): 158-79.

49. Grant, Verne. *The Origin of Adaptations.* New York: Columbia University Press, 1963.

50. Greene, John C. *The Death of Adam.* New York: New American Library, 1961.

51. Grene, Marjorie. *The Knower and the Known.* New York: Basic Books, 1966.

52. Haldane, J. S. *Mechanism, Life and Personality.* 2d ed. London: John Murray, 1921.

53. Haldane, J. S. *The Philosophical Basis of Biology.* New York: Doubleday, Doran & Co., 1931.

54. Hanson, N. R. *Patterns of Discovery.* Cambridge: Cambridge University Press, 1958.

55. Harré, R. *Matter and Method.* New York: St. Martin's Press, 1964.

56. Harris, E. E. *The Foundations of Metaphysics in Science.* New York: Humanities Press, 1965.

57. Hayek, F. A. "Theory of Complex Phenomena." In Mario Bunge, ed. *The Critical Approach to Science and Philosophy* (Glencoe, Ill.: The Free Press, 1964), pp. 332-49.

58. Hein, Hilde. "Molecular Biology vs. Organicism: The Enduring Dispute between Mechanism and Vitalism," *Synthese* 20 (1969): 238-53.

59. Hempel, Carl G. *Aspects of Scientific Explanation.* New York: The Free Press, 1965.

60. Hempel, Carl G., and Oppenheim, Paul. "Studies in the Logic of Explanation," *Philosophy of Science* 15 (1948): 135-75. Reprinted in *59*, pp. 245-90.

61. Hershey, A. D., and Chase M. "Independent Functions of Viral Protein and Nucleic Acid in Growth of Bacteriophage," *Journal of General Physiology* 36 (1952): 39-56.

62. Hess, Eckhard H. "Ethology," in *New Directions in Psychology* (New York: Holt, Rinehart and Winston, 1962), 1:157-266.

63. Hesse, Mary. *Models and Analogies in Science.* Notre Dame, Indiana: University of Notre Dame Press, 1966.

64. Hull, David. "Morphospecies and Biospecies: A Reply to Ruse," *British Journal for the Philosophy of Science* 21 (1970): 280-82.

65. Hull, David. "What the Philosophy of Biology Is Not," *Synthese* 20 (1969): 157-84.

66. Huxley, Julian S. *Evolution in Action.* New York: Harper, 1953.
67. Huxley, Julian S. *Evolutionary Ethics.* London: Oxford University Press, 1943.
68. Huxley, T. H. *Evolution and Ethics and Other Essays.* New York: D. Appleton & Co., 1929.
69. Jukes, Thomas H. *Molecules and Evolution.* New York: Columbia University Press, 1966.
70. Kant, Immanuel. *Critique of Teleological Judgement.* Translated by J. C. Meredith. Oxford: The Clarendon Press, 1928.
71. Kant, Immanuel. *First Introduction to the Critique of Judgment.* Translated by James Haden. Indianapolis: Bobbs-Merrill, 1965.
72. Kemeny, John G., and Oppenheim, Paul. "On Reduction," *Philosophical Studies* 7 (1956): 6-19.
73. King, Jack L., and Jukes, Thomas H. "Non-Darwinian Evolution," *Science* 164 (1969): 788-98.
74. Kornberg, Arthur. "Biologic Synthesis of Deoxyribonucleic Acid," *Science* 131 (1960): 1503-08.
75. Kuhn, Thomas. *The Structure of Scientific Revolutions.* Chicago: University of Chicago Press, 1962.
76. Lacey, James C., and Pruitt, Kenneth M. "The Origin of the Genetic Code," *Nature* 223 (1969): 779-804.
77. Lehman, Hugh. "Are Biological Species Real?" *Philosophy of Science* 34 (1967): 157-67.
78. Lehman, Hugh. "Functional Explanation in Biology," *Philosophy of Science* 32 (1965): 1-20.
79. Lehman, Hugh. "On the Form of Explanation in Evolutionary Biology," *Theoria* 32 (1966): 14-24.
80. Lehman, Hugh. "Reply to Munson," *Philosophy of Science* 37 (1970): 125-30.
81. Lenneberg, Eric H. *Biological Foundations of Language.* New York: John Wiley & Sons, 1967.
82. Lorenz, Konrad. *Evolution and the Modification of Behavior.* Chicago: University of Chicago Press, 1965.
83. Lorenz, Konrad. *On Aggression.* Translated by Marjorie Kerr Wilson. New York: Bantam Books, 1967.
84. Manser, A. R. "The Concept of Evolution," *Philosophy* 40 (1965): 18-34.
85. Maupertuis, Pierre-Louis de. *The Earthly Venus.* Translated by Simone Brangier Boas. New York: Johnson Reprint Corp., 1966.
86. Mayr, Ernst. *Animal Species and Evolution.* Cambridge, Mass.: Harvard University Press, 1963.

87. McMullin, Ernan. "What Do Physical Models Tell Us?" In B. van Rootselaar and J. F. Stahl, eds. *Logic, Methodology and Philosophy of Science III* (Amsterdam: North-Holland Publishing Co., 1968), pp. 385-96.

88. Medawar, P. B. *The Art of the Soluble.* London: Methuen, 1967.

89. Mendel, Gregor. "Experiments in Plant Hybridization." Translated by the Royal Horticultural Society, with modifications and comments by William Bateson. Cambridge: Harvard University Press, 1965.

90. Meselson, M., and Stahl, F. W. "The Replication of DNA in *Escherichia coli,*" Proceedings of the National Academy of Sciences 44 (1958): 671-82.

91. Moorhead, Paul S., and Kaplan, Martin M. eds. *Mathematical Challenges to the Neo-Darwinian Interpretation of Evolution.* Philadelphia: The Wistar Institute Press, 1967.

92. Morgan, T. H. "Chromosomes and Heredity," *American Naturalist* 44 (1910): 449-96.

93. Morgan, T. H. "Random Segregation versus Coupling in Mendelian Inheritance," *Science* 34 (1911): 384.

94. Morgan, T. H. "Sex Limited Inheritance in Drosophila," *Science* 32 (1910): 120-22.

95. Morgan, T. H. *The Theory of the Gene.* New Haven: Yale University Press, 1928.

96. Morris, Desmond. *The Naked Ape.* New York: Dell Publishing Co., 1969.

97. Muller, H. J. "Variation Due to Change in the Individual Gene," *American Naturalist* 56 (1922): 32-50.

98. Munson, Ronald. "Biological Species: Lehman's Thesis," *Philosophy of Science* 37 (1970): 121-25.

99. Nagel, Ernest. *The Structure of Science.* London: Routledge & Kegan Paul, 1961.

100. Olby, Robert C. *Origins of Mendelism.* London: Constable & Co., 1966.

101. Oparin, A. I. *Life: Its Nature, Origin and Development.* Translated from the Russian by Ann Synge. New York: Academic Press, 1961.

102. Orgel, L. E. "Evolution of the Genetic Apparatus," *Journal of Molecular Biology* 38 (1968): 381-93.

103. Pantin, C. F. A. *The Relations between the Sciences.* London: Cambridge University Press, 1968.

104. Pap, Arthur. *An Introduction to the Philosophy of Science.* New York: The Free Press of Glencoe, 1962.

105. Peirce, C. S. *Collected Papers,* vol. 5. Edited by Charles

Hartshorne and Paul Weiss. Cambridge, Mass.: Harvard University Press, 1934.

106. Polanyi, Michael. "Life's Irreducible Structure," *Science* 160 (1968): 1308-12.

107. Polanyi, Michael. *Personal Knowledge.* Harper & Row, 1964.

108. Polanyi, Michael. *The Tacit Dimension.* New York: Doubleday & Co., 1966.

109. Punnett, R. C. "Reduplication Series in Sweet Peas," *Journal of Genetics* 3 (1913): 77-103.

110. Quine, W. V. *From a Logical Point of View.* Cambridge, Mass.: Harvard University Press, 1953.

111. Rashevsky, N. "A Unified Approach to Biological and Social Organisms." In B. van Rootselaar and J. F. Stahl, eds. *Logic, Methodology and Philosophy of Science III* (Amsterdam: North-Holland Publishing Co., 1968), pp. 403-12.

112. Ravin, Arnold W. *The Evolution of Genetics.* New York: Academic Press, 1965.

113. Roll-Hansen, Nils. "On the Reduction of Biology to Physical Science," *Synthese* 20 (1969): 277-89.

114. Ruse, Michael. "Definitions of Species in Biology," *British Journal for the Philosophy of Science* 20 (1969): 97-119.

115. Russell, Bertrand. "On the Notion of Cause," in his *Mysticism and Logic* (London: George Allen & Unwin, 1917), pp. 180-208.

116. Russell, E. S. *The Interpretation of Development and Heredity.* Oxford: The Clarendon Press, 1930.

117. Schaffner, Kenneth F. "Antireductionism and Molecular Biology," *Science* 157 (1967): 644-47.

118. Schaffner, Kenneth F. "Approaches to Reduction," *Philosophy of Science* 34 (1967): 137-47.

119. Schaffner, Kenneth F. "Theories and Explanations in Biology," *Journal of the History of Biology* 2 (1969): 19-33.

120. Schaffner, Kenneth F. "The Watson-Crick Model and Reductionism," *British Journal for the Philosophy of Science* 20 (1969): 325-48.

121. Scriven, Michael. "Explanation and Prediction in Evolutionary Theory," *Science* 130 (1959): 477-82.

122. Scriven, Michael. "Explanation in the Biological Sciences," *Journal of the History of Biology* 2 (1969): 187-98.

123. Scriven, Michael. "Explanations, Predictions, and Laws." In H. Feigl and G. Maxwell, eds. *Minnesota Studies in the Philosophy of Science,* vol. 3 (Minneapolis: University of Minnesota Press, 1962), pp. 170-230.

124. Scriven, Michael. "Truisms as the Grounds for Historical

Explanation." In Patrick Gardiner, ed. *Theories of History* (Glencoe, Ill.: The Free Press, 1959), pp. 443-75.

125. Shapere, Dudley, "Biology and the Unity of Science," *Journal of the History of Biology* 2 (1969): 3-18.

126. Simon, Herbert A. *The Sciences of the Artificial.* Cambridge, Mass.: MIT Press, 1969.

127. Simon, Michael A. "Could There Be a Conscious Automaton?" *American Philosophical Quarterly* 6 (1969): 71-78.

128. Simon, Michael A. "Materialism, Mental Language, and Mind-Body Identity," *Philosophy and Phenomenological Research* 30 (1970): 514-32.

129. Simpson, George Gaylord. *Biology and Man.* New York: Harcourt, Brace & World, 1969.

130. Simpson, George Gaylord. *The Meaning of Evolution.* 2d rev. ed. New Haven: Yale University Press, 1967.

131. Simpson, George Gaylord. *This View of Life.* New York: Harcourt, Brace & World, 1964.

132. Singer, Charles. *A History of Biology.* 3d ed. London: Abelard-Schuman, 1959.

133. Sinnott, Edmund W. *Matter, Mind and Man.* New York: Atheneum, 1962.

134. Smart, J. J. C. *Between Science and Philosophy.* New York: Random House, 1968.

135. Smart, J. J. C. *Philosophy and Scientific Realism.* London: Routledge & Kegan Paul, 1963.

136. Spector, Marshall. "Models and Theories," *British Journal for the Philosophy of Science* 16 (1965): 121-42.

137. Spencer, Herbert. *The Principles of Biology.* 2 vols. New York: D. Appleton & Co., 1864.

138. Stent, Gunther S. "That Was the Molecular Biology That Was," *Science* 160 (1968): 390-95.

139. Sturtevant, A. H. "The Himalayan Rabbit Case, with Some Considerations on Multiple Allelomorphs," *American Naturalist* 47 (1913): 234-38.

140. Sturtevant, A. H. "The Reduplication Hypothesis as Applied to Drosophila," *American Naturalist* 48 (1914): 535-49.

141. Sutton, Walter S. "The Chromosomes in Heredity," *Biological Bulletin* 4 (1903): 231-51.

142. Swanson, J. W. "On Models," *British Journal for the Philosophy of Science* 17 (1967): 297-311.

143. Taylor, Charles. *The Explanation of Behaviour.* London: Routledge & Kegan Paul, 1964.

144. Taylor, J. H.; Woods, P. S.; and Hughs, W. L. "The Organi-

zation and Duplication of Chromosomes as Revealed by Autoradiographic Studies Using Tritium-labeled Thymidine," *Proceedings of the National Academy of Sciences* 43 (1957): 122-28.

145. Teilhard de Chardin, Pierre. *The Phenomenon of Man.* Translated by Bernard Wall. London: Wm. Collins Sons & Co., 1959.

146. Thorpe, W. H. *Science, Men and Morals.* Ithaca, N.Y.: Cornell University Press, 1965.

147. Tinbergen, N. *Social Behavior in Animals.* London: Methuen, 1953.

148. Toulmin, Stephen. *Foresight and Understanding.* New York: Harper & Row, 1963.

149. Toulmin, Stephen. *Philosophy of Science.* New York: Harper & Row, 1960.

150. Trow, A. H. "Forms of Reduplication—Primary and Secondary," *Journal of Genetics* 2 (1913): 313-24.

151. Vries, Hugo de. *Intracellular Pangenesis.* Translated by C. Stuart Gager. Chicago: Open Court Publishing Co., 1910.

152. Vries, Hugo de. "The Law of Segregation in Hybrids." Translated by Evelyn Stern, in Curt Stern and Eva R. Sherwood, eds. *The Origin of Genetics* (San Francisco: W. H. Freeman & Co., 1966), pp. 107-17.

153. Waddington, C. H. *The Ethical Animal.* Chicago: University of Chicago Press, 1960.

154. Waddington, C. H. *The Nature of Life.* New York: Harper & Row, 1966.

155. Waddington, C. H., ed. *Towards a Theoretical Biology*, vols. 1-2. Chicago: Aldine Publishing Co. 1968, 1969.

156. Watson, James D. *The Double Helix.* New York: Atheneum, 1968.

157. Watson, James D., and Crick, F. H. C. "Genetical Implications of the Structure of Deoxyribonucleic Acid," *Nature* 171 (1953): 964-67.

158. Watson, James D., and Crick, F. H. C. "Molecular Structure of Nucleic Acids," *Nature* 171 (1953): 737-38.

159. Weismann, August. "On the Number of Polar Bodies, etc." In *Essays upon Heredity and Kindred Biological Problems.* Translated by Edward B. Poulton, Selmar Schönland, and Arthur E. Shipley (Oxford: The Clarendon Press, 1889), pp. 333-84.

160. Whitehead, Alfred North. *Science and the Modern World.* New York: The Macmillan Co., 1925.

161. Whitehouse, H. L. K. *Towards an Understanding of the*

Mechanism of Heredity. New York: St. Martin's Press, 1967.

162. Wigner, Eugene P. "The Probability of the Existence of a Self-Reproducing Unit." In *The Logic of Personal Knowledge* (London: Routledge & Kegan Paul, 1961), pp. 231-38.

163. Wittgenstein, Ludwig. *Philosophical Investigations.* Translated by G. E. M. Anscombe. New York: The Macmillan Co., 1953.

164. Woese, Carl R. *The Genetic Code.* New York: Harper & Row, 1967.

165. Woese, Carl R. "Models for the Evolution of Codon Assignments," *Journal of Molecular Biology* 43 (1969): 235-40.

166. Woodger, J. H. *The Axiomatic Method in Biology.* Cambridge: Cambridge University Press, 1937.

167. Woodger, J. H. *Biological Principles.* Reissued, with a new Introduction. London: Routledge & Kegan Paul, 1967.

168. Zirkle, Conway. "Gregor Mendel and His Precursors," *Isis* 42 (1951): 97-104.

Index